Nilufarhan Hamrabaeva

Seismic risk assessment of hazardous production facilities

Nilufarhan Hamrabaeva

Seismic risk assessment of hazardous production facilities

Methodological approach and technical solutions

ScienciaScripts

Imprint

Cover image: www.ingimage.com

This book is a translation from the original published under ISBN 978-620-5-51383-5.

Publisher:
Sciencia Scripts
is a trademark of
Dodo Books Indian Ocean Ltd. and OmniScriptum S.R.L publishing group

120 High Road, East Finchley, London, N2 9ED, United Kingdom
Str. Armeneasca 28/1, office 1, Chisinau MD-2012, Republic of Moldova, Europe
Printed at: see last page
ISBN: 978-620-5-64259-7

Ministry of Higher and Secondary Education
Tashkent Architecture and Construction Institute

Khamrabayeva Nilufarkhan Azizovna

MONOGRAPHY

Development of a methodological approach and technical solutions for seismic risk assessment of hazardous production facilities

Tashkent - 2022

UDC 622.1.\7

LBC 68.53

Approved by TASI Scientific Council Decision No. 1 of 10.09.2022

Khamrabaeva N.A. Development of methodological approach and technical solutions for seismic risk assessment of hazardous production facilities. Monograph. - Saarbrücken (Germany). Lambert Academic Publishing, TASI, 2022, - 111 p.

The monograph is designed to improve the effectiveness of risk assessment and contains methods of risk analysis and risk assessment applicable to various hazardous production facilities, as well as examples of the use of the approaches and methods considered.

The monograph focuses on the risk management process, in the development of risk assessment documents, as well as in the development of standards, specifications and other regulatory documents for risk management at hazardous production facilities.

Specialties:

05.10.02: Safety in emergency situations. Fire, Industrial, Nuclear and Radiation Safety.

Direction:

05.10.02: Safety in emergency situations. Fire, industrial, nuclear and radiation safety.

Keywords

Safety hazards, process, risks at industrial facilities, hazards, safety assurance, safety mechanisms and techniques, seismic effects, seismic safety.

Table of contents

RECENTLY

Khamrabayeva Nilufarkhan Azizovna's monograph "Development of a methodological approach and technical solutions for assessing the seismic risk of hazardous production facilities

State policy in the field of environmental and industrial safety and new concepts for ensuring safety and accident-free production processes at economic facilities, dictated by the Law "On Industrial Safety of Hazardous Production Facilities" dated 28.09.2006, No. No. ZRU -57, Law No. 120-II of August 31, 2000 on Radiation Safety of Population, Law No. ZRU- 393 of August 26, 2015 on Sanitary and Epidemiological Welfare of Population, Law No. 754 -XII of December 9, 1992 on Use of Atomic Energy primarily provides for objective assessment of hazards and allows to outline ways to combat them.

Environmental and technological safety is a state of reality in which the occurrence of a hazard is excluded with a certain probability. A hazardous situation occurs when a person is in a hazardous area, i.e. in a space where hazards due to hazardous or harmful factors occur continuously, periodically or episodically. Dangerous situations are realised due to a combination of reasons causing the impact of hazardous or/and harmful factors on a person, resulting in gradual or instantaneous damage to his/her health.

The material presented in this monograph is timely and will be useful to scholars working in the field of complex system security. It is recommended that this monograph be published in the public domain.

Reviewer:

Professor, Tashkent University
state technical
University Dr. of Technology, Professor, Department of
"Life safety"
I.A. Suleymanov, A.A. Karimov I.A. Suleimanov A.A.

RECENTLY

Khamrabayeva Nilufarkhan Azizovna'**s monograph** "Development of a methodological approach and technical solutions for assessing the seismic risk of hazardous production facilities

According to the UN Secretary-General, the damage caused by man-made disasters has tripled over the past 30 years, reaching $200 billion a year. In Russia, cumulative annual material damage caused by man-made disasters is estimated at USD 200 billion per year. In Russia the total annual material damage from man-made accidents, including costs of their elimination, exceeds 40 billion roubles. Emergency (ES) is a set of events and hazards, suddenly violating existing conditions of life, threatening life and health of people, their environment, elements of technosphere. Technogenic emergency (technogenic ES) - a state when as a result of occurrence of
technogenic emergency - a condition resulting in emergence of a source of technogenic emergency at a facility, certain territory or water area which disturbs normal conditions of life and activity of people, endangers their life and health, damages property of population, national economy and natural environment.

Each emergency can be seen as a large-scale hazardous situation that simultaneously threatens a large number of people and technosphere facilities. The stages of origin and development of an emergency are usually latent and are associated with an accumulation of destructive potential. In the culminating stage, a multitude of hazardous and damaging factors combine to form one or more damaging factors.

Emergencies (ES) arise from both natural and man-made accidents. Emergency situations are most common in the coal, mining, chemical, oil and gas, metallurgical and transport industries. Emergencies in industrial and domestic environments are often associated with depressurization of pressurized systems (cylinders and tanks for storage or transportation of compressed, liquefied and dissolved gases, gas and water pipelines, heat supply systems, etc.).

The monograph is intended for workers in the "Life Safety" sector, as well as university lecturers, masters, experts, safety engineers and researchers. Therefore, this monograph should be published in the public domain.

Reviewer:

Professor, Tashkent Architectural and Engineering University of the Institute of Civil Engineering Doctor of Chemistry, Professor

of the Department of Construction Materials and Chemistry
Mukhamedgaliev B.A.

Introduction

Risk management is the process of adopting and implementing management decisions aimed at reducing the likelihood of an adverse outcome and minimising the potential losses caused by its realisation.

The main objective of infrastructure risk management is to achieve and maintain an acceptable level of safety at hazardous production facilities.

Risk management at a hazardous production facility (HPF) must be carried out in strict accordance with the requirements of national risk management standards.

A methodology should be used to effectively manage the risks in HSE, which aims to

- to identify the risk and assess the likelihood of its occurrence and the scale of its consequences;
- to determine the maximum possible loss;
- on the choice of methods and tools for managing the identified risk;
- to develop a risk management strategy to reduce the likelihood of risk materialising and to minimise the possible negative consequences;
- on the implementation of the risk management strategy;
- to assess the results achieved and to adjust the risk management strategy.

When developing and implementing concepts, strategies, programmes and regulations for the technical and technological development of a hazardous production facility, it is useful to consider two main aspects:

- predictive efficiency of railway transport operations at all stages of the life cycle;
- the associated operational risk calculations.

The actual performance will be determined by the difference between the expected economic effect resulting from the implementation of planned activities and possible losses (risks). This approach to quantitative assessment of technical and technological development of OSR has not been systematically used before.

Each of the stages of adverse processes can be characterised by risks as quantitative indicators of the possible manifestation of hazards.

Thus, taking into account possible risks, the main programmes and activities undertaken (formation of strategies, concepts, reform plans, infrastructure development) should be assessed, both the level of positive effects as well as the level of risks for each stage of implementation of the planned activities.

Obviously, risks are inherent not only in human life, but also in the activities of organisations, businesses, territories, buildings - any processes and systems. However, unlike a thinking person, all of these entities do not have self-regulating mechanisms, reason and common sense, and, in addition, the more complex the system or process, the more risks are inherent in it, and the more difficult it is to build a scheme to manage such risks. An adult, as an individual member of the homo sapiens species, is largely responsible for himself, but if he is the head of a large organisational structure or a specialist responsible for the operation of a system or process, if he has a duty to make decisions that are not his alone, then he cannot avoid managing the risks involved.

CHAPTER 1. METHODOLOGY FOR RISK ASSESSMENT OF HAZARDOUS PRODUCTION FACILITIES AND INFRASTRUCTURE

1.1 Risk management process methodology

In dealing with complex safety issues in developed countries, a risk management process methodology is widely used, based on determining the frequency (likelihood) and consequences of undesirable events.

The combination of two conditions - the possibility of the occurrence of an undesirable event and the susceptibility of an object to its influence - is a sufficient basis for recognising the existence of risk.

Based on a review of risk management research, taking into account current requirements, the risk management methodology should meet the following principles:

– the risk decision must be economically sound and must not have a negative impact on the financial and economic performance of the railway transport organisation;

– risks should be managed as part of the corporate strategy of the hazardous production facility (HPF);

– In risk management, decisions must be based on the right amount of reliable information;

– when managing risks, the decisions taken should take into account the objective characteristics of the environment in which the railway transport organisation operates;

– Risk management must be systematic;

– risk management should involve ongoing review of the effectiveness of the decisions taken and prompt adjustment of the set of risk management principles and methods used.

The risk management process [1] is illustrated in Figure 1.1.

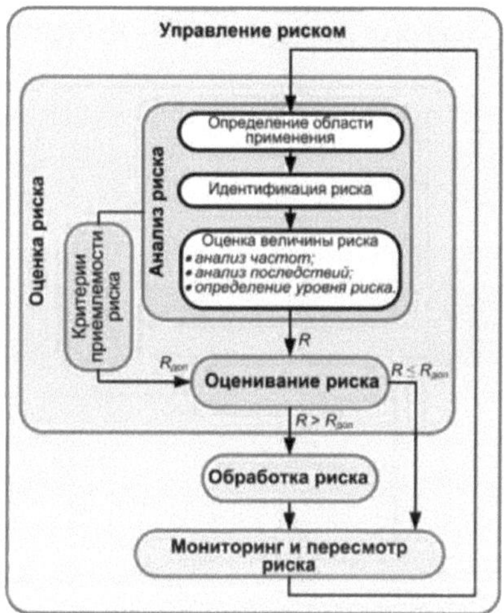

where R is the level of risk, $R_{доп}$ is the acceptable level of risk.

Figure 1.1 - Risk management process

It is important to document the individual stages of the risk management process. It is important to document the risk assessment phase, usually in the form of a risk assessment report. Documenting the risk assessment allows the accumulated information to be used later for risk treatment, risk reduction costing, and a number of other factors.

The essence of each stage of risk management involves the application of different methods. These methods are organised into a step-by-step process for implementing risk management.

The methodological principles for organising the risk management process are based on the sequential implementation of individual steps, procedures and procedures, using approaches, methods, technologies known and applied in practice, available information on external and internal conditions and the object or process to be managed.

A breakdown of the risk management process into building blocks (steps, procedures and processes), according to Figure 1.1, is shown in Table 1.1.

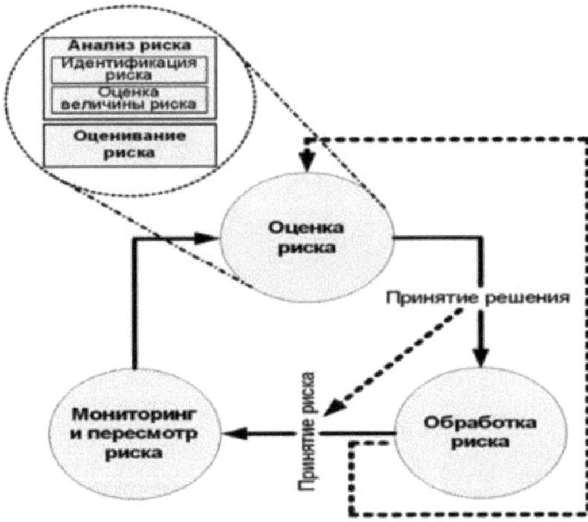

Figure 1.1. - Subdividing the risk management process into building blocks

In the risk assessment phase, hazards are identified, the magnitude of the risk is assessed and compared (when assessed) to established thresholds defined by risk criteria. After the risk assessment, a decision is made whether or not to accept the risk. If the risk is not accepted, treatment is carried out to reduce it. If the residual risk value (after treatment) cannot be accepted, the "risk assessment" and "risk treatment" steps (or just the "risk treatment" step) are repeated.

Table 1.1 Structural elements of the risk management process

<table>
<tr><th></th><th>Stages</th><th>Procedures</th><th colspan="2">Steps</th></tr>
<tr><td rowspan="8">Risk management</td><td rowspan="6">1. risk assessment</td><td rowspan="5">1.1 Risk analysis</td><td colspan="2">1.1.1 Defining the field of application</td></tr>
<tr><td colspan="2">1.1.2 Risk identification</td></tr>
<tr><td rowspan="3">1.1.3 Risk assessment</td><td>a) Frequency analysis</td></tr>
<tr><td>b) Impact analysis</td></tr>
<tr><td>c) Determining the level of risk</td></tr>
<tr><td>1.2 Risk assessment</td><td></td><td></td></tr>
<tr><td>2. handling the risk</td><td></td><td></td><td></td></tr>
<tr><td>3. risk monitoring and review</td><td></td><td></td><td></td></tr>
</table>

The relationship between the main stages of risk management is illustrated in Figure 1.2.

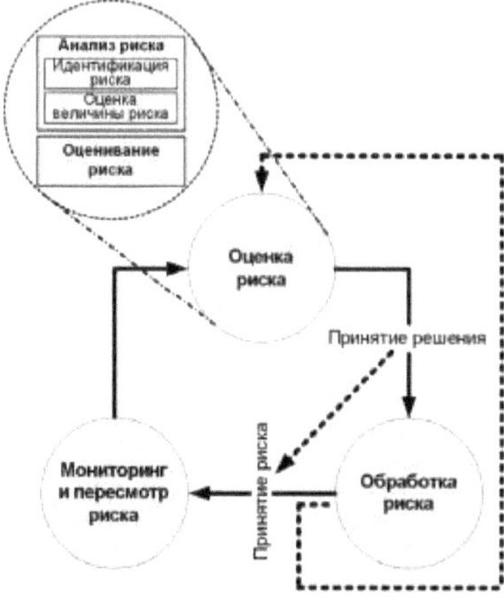

Figure 1.2 - Relationship between the stages of risk management

The basic step to shape the risk management strategy is **the risk assessment phase**.

The main part of the risk assessment stage is the **risk analysis** procedure, which has a special place in the risk management process and determines the effectiveness of risk reduction.

The schematic link between risk analysis and risk reduction is quite simple and is shown in Figure 1.3.

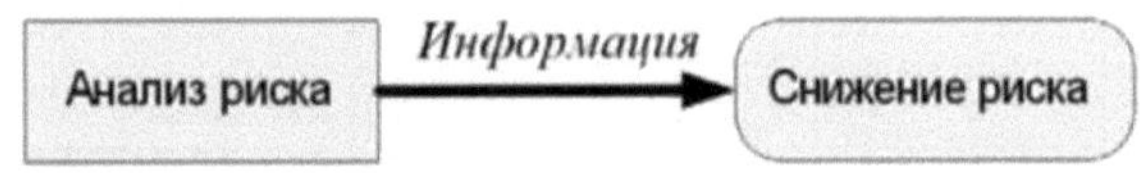

Figure 1.3 - Relationship between analysis and risk reduction

At different stages of the infrastructure and rolling stock lifecycle, the specific objectives of the risk analysis may vary. In the pre-design or design phase, the purpose of the risk analysis may be:

– identifying hazards and assessing the magnitude of the risk, taking into account the impact of influencing factors on personnel, the public, physical facilities and the environment;

– Consideration of the results in analysing the acceptability of the proposed solutions and selecting the best options for the location of the equipment, the facility, taking into account the characteristics of the environment;

– providing information for the development of instructions, procedures and plans for dealing with hazardous situations;

– assessment of different options for design and development proposals. In the operational and refurbishment phase, the purpose of the risk analysis may be

– comparison of the operating conditions of the facility with the relevant safety requirements;

– clarification of information on the main dangers;

– Making recommendations to justify or modify regulatory requirements, licensing issues, determining the frequency of condition inspections, safety inspections, etc;

– Improvement of operation and maintenance manuals, hazard localisation plans;

– Assessing the effect of changes in organisational structure, practices and maintenance on safety performance.

During the decommissioning (or commissioning) phase, the purpose of the risk analysis may be

– identifying hazards and assessing their consequences;

– providing information to develop or refine decommissioning (or commissioning) instructions.

In order to identify hazards and assess their consequences, a decomposition of the hazardous situation is usually carried out. An example of a decomposition is shown in Figure 1.4.

Figure 1.4 - Decomposition of a hazardous situation

Hazard decomposition assesses the extent and nature of the impact of each hazard on people, the environment, infrastructure and rolling stock as well as on processes.

The company's activities should be carried out in such a way that the possible risks in their implementation would be below the level of acceptable risks, which are rationed in advance based on the overall concept (purpose) of the specific activities, including the expected final economic effect.

The levels of acceptable risks are determined and assigned by the HSE management bodies taking into account scientific, technical and economic capabilities. Optimal planning of technical and technological development of EP is to ensure the greatest excess of benefits (positive result) from the planned activities over the possible damages in their implementation.

When assessing the effectiveness of planned activities against risk parameters, it is necessary to

– forecasting of adverse events, scientific justification of risk assessments in the implementation of planned activities, both individually and holistically;

– scientific justification of critical (unacceptable) risk assessments;

– setting levels of acceptable risks, which are decided upon by management bodies in an expert or decision-making manner, based on the task at hand;

– managing technical and technological development with safety in mind according to risk criteria.

The ultimate goal of implementing the methodology under consideration is to create an opportunity to forecast, plan and manage company development using new approaches to the quantitative assessment of differentiated (for individual facilities of the HSE), integral (for the basic structures of HSE), complex (for the reform and functioning of HSE) and strategic risks.

The analysis of risks during the implementation of activities, their assessment and comparison with permissible levels will make it possible to reasonably decide, using quantitative indicators, on the permissibility or, on the contrary, on the inadmissibility of the implementation of certain projects, on the direction of their revisions and adjustments leading to risk mitigation.

Universal indicators to express differentiated or integral, integrated and strategic risks in practice are:

– harm to the life and health of participants in the transport process and third parties;

– the economic equivalent of adverse events at hazardous sites.

In order to improve the safety and operational efficiency of HSE infrastructure, specific risk mitigation programmes with specific costs depending on the magnitude of the risks should be envisaged.

Both common scenarios (optimistic, inertial and pessimistic) and the following scenarios should be considered in the functional safety analysis of infrastructure and rolling stock:

– operation of the facility under normal conditions (normal situations);

– the operation of the facility with acceptable deviations from normal conditions with a return to the initial condition;

– The operation of a facility with deviations from normal conditions caused by adverse events and requiring special measures to return to its original state;

– functioning of the facility with deviations from normal conditions caused by "out-of-design situations", when decisions made, activities or processes have to be revised, and the facility has to be switched to a new state with a given level of vulnerability (resistance to the impact of adverse events);

– functioning of the facility under hypothetical conditions with adverse events developing under worst (severe) implementation scenarios with unforeseen triggering factors preventing the implementation of activities or processes related to the functioning of the facility.

According to the legislation, infrastructure facilities are conditionally divided into the following categories according to their level of risk, technical complexity, potential hazard and functional significance:

– objects of technical regulation; - hazardous production facilities;

– critical facilities.

In all cases of risk analysis for each of the above categories of facilities, a three-component system of their interaction with external environment factors should be assumed:

– the social factor (human interaction);

– technogenic factor (interaction with the production process);

– the natural factor (interaction with the environment).

In practice, in risk assessment, the realisation of hazards is often assessed by the uncontrolled release (spread) of major lethal factors, which include energy, matter, information and others.

The following will address the stages, procedures and steps of the risk management process and their corresponding methods. The objectives of the risk assessment are:

– obtaining reliable baseline information; - carrying out the necessary analysis;

– making informed decisions when assessing risk;

– forming the baseline data for further selection of optimal risk treatment solutions.

The risk assessment phase includes the following procedures:

– risk analysis;

– risk assessment.

Risk analysis, in turn, consists of the following steps:

– definition of the area of application;

– risk identification;

– an assessment of the magnitude of the risk.

Assessing the magnitude of the risk involves analysing frequencies (or probabilities), analysing consequences and determining the level of risk.

Also at the risk assessment stage, acceptable risk criteria may be established if they have not been previously defined (e.g. strictly set out in regulatory documentation).

In the process of defining the scope of risk analysis, it is necessary to

– state the reasons and problems that have led to the need for a risk analysis;

– Define the subject of the risk assessment and describe it;

– select appropriate specialists to carry out the risk analysis;

– establish the sources of information on the safety of the risk assessment object;

– state the background data and constraints that condition the scope of the risk assessment;

– clearly define the aims and objectives of the risk analysis;

– justify the risk analysis methods used;

– define in advance the criteria for acceptable risk.

The definition of the scope of risk analysis is carried out by examining the properties of the risk assessment object - human, environmental or infrastructure and rolling stock, analysis of the main indicators of the object, internal and external conditions, as well as other input data. The main purpose of studying the subject of risk assessment is to identify sources and methods of using information about the subject.

Each undesirable event can occur in relation to a particular risk object. A distinction is made between individual, technical, environmental, social and economic risks. Each type is driven by characteristic sources and factors of risk, the classification and characterisation of which are shown in Table 1.2.

Table 1.2 - Classification and characterisation of risks

Type of risk	**Risk exposure**	**Source of risk**	**An undesirable event**
Individual	The man	The human condition	Illness, injury, disability, death
Technical	Technical systems and facilities	Technical inadequacies, violation of the rules of operation of technical systems and facilities	Accident, explosion, disaster, fire, destruction
Environmental	Environmental Systems	Anthropogenic interference with the natural environment, man-made emergencies	Anthropogenic environmental disasters, natural disasters
Social	Social groups	Emergency, reduced quality of life	Group injuries, diseases, fatalities, increased deaths
Economic	Material resources	Increased production or environmental hazards	Increased security costs, damage from insufficient security

In determining the scope of the risk assessment, the following steps are carried out:

(a) A description of the reasons and issues that prompted the risk analysis, including

1) formulating the goals and objectives of the risk analysis;

2) Definition of failure criteria for infrastructure and rolling stock;

b) a description of the risk assessment object (person, environment, infrastructure and rolling stock) is drawn up, including

1) A general description of the subject of the risk assessment;

2) Definition of boundaries and areas of contact with adjacent infrastructure and rolling stock;

3) a description of the external conditions (environmental conditions);

4) Identification of internal conditions (operating conditions and states of the infrastructure and rolling stock to which the risk assessment applies and the associated constraints);

c) assumptions and assumptions are recorded;

d) the formulation of the decisions that can be taken, the description of the required outputs from the studies and from the decision-makers is developed.

1.2 Risk identification

The risk identification phase identifies a list of adverse events whose occurrence is, firstly, realistic and, secondly, likely to degrade the quality of the environment and cause harm to humans or the infrastructure and rolling stock.

The procedure for carrying out risk identification is shown in Figure 1.5.

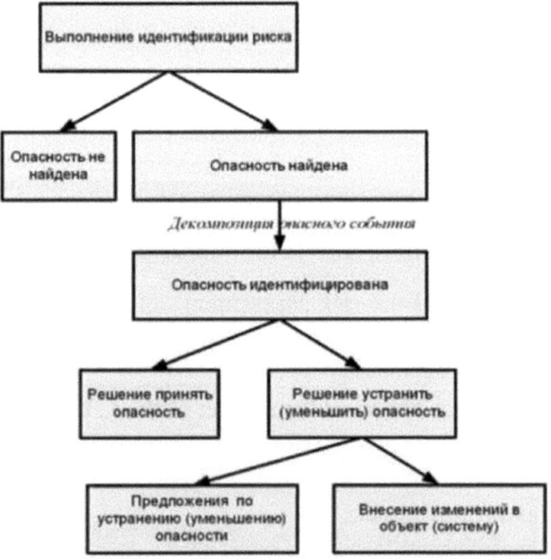

Figure 1.5 - Performing risk identification

In the initial identification phase, a preliminary hazard analysis is carried out to identify and describe hazardous systems, subsystems and components of the facility, as well as other sources of hazards.

The results of the preliminary hazard analysis and the application of hazard identification methods make it possible to determine which components, subsystems, systems or processes require more serious analysis and which are of less interest.

The result of risk identification is a list of undesirable events leading to an accident or other undesirable consequences. As a result of risk identification, further actions for risk analysis are determined:

– the decision to discontinue further risk analysis due to the insignificance of the hazards;

– the decision to carry out a more detailed risk analysis;

– making recommendations to reduce or eliminate hazards. Different methods can be used to identify risk: statistical, analytical, expert, analogy methods and others.

Risk assessment is the determination of the magnitude (measure) of the risk to human health, the environment and property in situations involving the realisation of a hazard. Risk assessment is a mandatory part of risk analysis and includes frequency analysis, analysis of consequences and their combinations, and determination of the risk level.

By definition, assessing the magnitude of the risk involves a frequency analysis and a consequence analysis. However, when the consequences are small and the frequency is extremely low, it is sufficient to estimate one of these parameters.

The purpose of the frequency analysis is to determine the frequency of occurrence of each of the undesirable events or accident scenarios identified at the risk identification stage.

The frequencies of events can be specified both qualitatively and quantitatively (as an interval of numerical frequency values).

The following methods are used to determine the frequency of occurrence of an event:

– estimating the frequency with which a given event has occurred in the past based on statistical data (data accumulated over a period of operation of the infrastructure or rolling stock in question, statistics on transport accidents and other events, etc.) and predicting the frequency with which the event will occur in the future;

– An estimate of the frequency of occurrence of a given event on the basis of data on failures of technical facilities occurring over a given period of time per unit of operational work for each railway transport economy;

– Predicting event frequencies using Failure Tree Analysis (FTA) and Event Tree Analysis (ETA);

– assessment based on expert judgement. Any available information on the infrastructure or rolling stock: statistical, experimental, design, etc., should be taken into account when carrying out expert assessments.

There are methods for obtaining expert assessments that eliminate ambiguity, such as the Delphi method, pairwise matching, risk group classification and others.

The resulting event frequency estimates are correlated with defined frequency levels. The number of frequency levels to be used and their characteristics are determined by the management of the railway transport organisation according to the specific conditions.

Typical levels of frequency (likelihood) of a hazardous event and a description of each level are shown in Table 1.3, where numerical expressions of frequencies are given as an example.

The frequency analysis produces data on the frequency of a hazardous event, ranked according to the frequency levels specified for that event.

Consequence analysis involves assessing the results of the impact of an undesirable event on people, property and the environment.

Consequence analysis can be performed either as a simple description of the results using simplified analytical methods or as detailed quantitative modelling (e.g. using computer simulation models).

Table 1.3. - Typical levels of event occurrence frequencies

Frequency level	**Frequency of events, *f*, year-1**	**Description**
Frequent	f > 10-3	The likelihood of frequent occurrence. Constant presence of danger
Probable	5×10-4 ≤ f < 10-3	Repeated occurrence. Frequent occurrence of a hazardous event is expected
Random	10-4 ≤ f < 5×10-4	Probability of repeated occurrence. Repeated occurrence of a hazardous event is expected
Rare	10-5 ≤ f < 10-4	The probability that an event will occasionally occur during the life cycle of a facility. The reasonable expectation that a hazardous event will occur.
Extremely rare	10-6 ≤ f < 10-5	The occurrence of an event is unlikely, but possible. It can be assumed that a dangerous situation could occur in an exceptional case.
The unlikely	f ≤ 10-6	The probability of occurrence is highly unlikely. It can be assumed that a hazardous event will not occur.

In carrying out an impact analysis, the following is carried out:

(a) Selection of a hazardous event based on the results of hazard identification;

b) a description of all consequences resulting from the hazardous event, including

1) the consequences that have caused damage to the risk assessment object in question;

2) consequences that may occur after a certain period of time (if considered within the scope of risk analysis);

3) secondary effects extending to adjacent infrastructure and rolling stock.

c) consideration of mitigation measures, taking into account all factors influencing the effects.

The resulting estimates of the consequences of a hazardous event are correlated with specified levels of consequence severity.

The number of consequence severity levels to be used and their characteristics are determined by the management of the railway transport organisation depending on the specific conditions.

Typical levels of consequence severity are shown in Table 1.4.

The consequence severity analysis produces data on the severity of the consequences of a hazardous event, ranked according to the consequence severity levels specified for that event.

The determination of the level of risk is made after the frequency analysis and the consequence analysis have been completed.

Table 1.4 - Typical levels of severity of event consequences

Levels of consequence severity	Consequences by type of risk	
	internal risks	**external risks**
Catastrophic	Death of 1 or more persons or serious injuries to 5 or more persons connected with railway operations **or** The rolling stock object is damaged to the extent of being removed from the inventory **or** Damage to infrastructure of more than 5,000 minimum wages	Death of 1 or more persons or serious injuries to 5 or more persons connected with railway operations **or** Damage to the environment causing a federal or interregional emergency
Critical	Serious injury to up to 5 persons in connection with railway operations. Death of 1 person or serious injury to 1 or more persons as a result of intentional or negligent acts of the victim or others not related to the operation of the HSE. **or** Damage to a rolling stock facility requiring major repairs to restore its serviceability **or** Damage to infrastructure of between 1,500 and 5,000 minimum wages **or** Total loss of cargo	Serious bodily injury to up to 5 persons connected with the operation of an HSE. Death or serious bodily injury to 1 or more persons as a result of the intentional or negligent actions of the victim himself or others not connected with the operation of the WGHE. **or** Damage to the environment causing a regional or inter-municipal emergency
Irrelevant	Damage to health of moderate severity **or** Damage to the rolling stock object requiring a medium or depot repair to restore it to serviceable condition **or** Damage to infrastructure of between 500 and 1,500 minimum wages **or** Partial loss of cargo	Damage to health of moderate severity **or** Damage to the environment causing a municipal or local emergency
Minor	Mild harm to health **or** Damage to a rolling stock facility requiring routine repairs to restore its serviceability **or** Damage to an infrastructure facility of less than 500 minimum wages	Mild harm to health **or** Minor damage to the environment
Note - The minimum wage is the minimum wage.		

In determining the level of risk, a quantitative assessment is carried out using various mathematical formulations.

$$R = P\text{-}C.$$

In general, the definition of risk level *R* involves expressing risk using two quantities - the probability *P* of an undesirable event and its consequences *C*. Often, in determining the level of risk, probability and consequence severity act as multipliers:

For unknown outcomes, risk can be expressed in terms of probability or frequency. The definition of risk as the probability of some random parameter exceeding a certain limit (threshold) can also be used.

Figure 1.6 shows a representation of the risks of relevant adverse events and their measures.

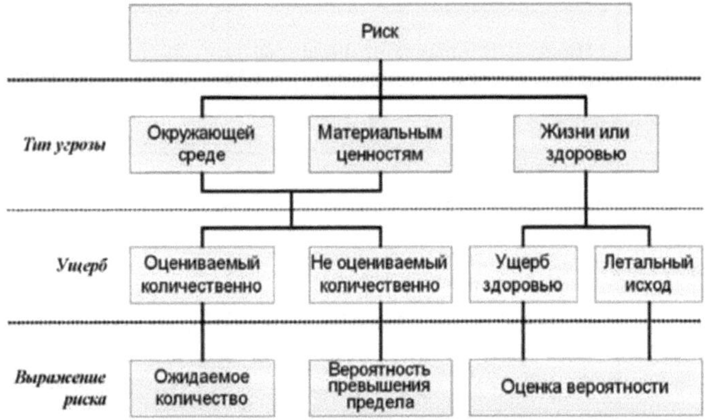

Figure 1.6 - Representation and risk measures

In the case of material hazards, risk is often measured in monetary terms. If the various consequences of an undesirable event are the same or very large, it is sufficient to consider only the corresponding probabilities for comparison. Alongside this there may be hazards that cannot be quantified, for example when the consequences of an event cannot be sufficiently foreseen. An example would be the consequences of the failure of a component used in different sectors of the economy, which the supplier of the component cannot estimate. In such a case, the measure of risk has to be taken as the probability of exceeding the load limit of the system in which the component was operated.

For health risks, consequences can be partly quantified in categories such as downtime or the cost of replacement staff, insurance benefits, etc. In the case of a fatal risk, quantification of the consequences is largely unavailable.

Particular problems arise when hazards threaten people, the environment and material assets at the same time. In such cases, it is advisable to assess the measure of risk against more than one criterion.

The risk assessment results in a quantitative (semi-quantitative or qualitative) expression of risk, which is a generalised characteristic of that risk and is used in the subsequent stages of the risk management process.

Risk tolerance criteria determine the acceptable level of risk and are set according to the methods of risk analysis, availability of necessary information, capacity and the objectives of the analysis. The risk tolerance criteria may include:

– set by the legal and regulatory framework;

– to be determined when planning the risk analysis;

– be determined in the process of obtaining the results of the risk analysis.

The main requirements for selecting an acceptable risk criterion in a risk analysis are that it is reasonable and certain.

The basis for determining acceptable risk should generally be

– legislation of the Republic of Uzbekistan;

– the safety rules and regulations in force in the area analysed;
– additional requirements of the responsible security authorities;
– information on existing emergency events and their consequences;
– experience in this type of activity.

Criteria for acceptable risk are usually based on operational, technical, economic, regulatory, social or environmental factors, or a combination of these. The basic principles of risk acceptance (ALARP, MEM, GAMAB) are discussed in [2].

The acceptable level of risk can be determined, for example, on the basis of available statistical data using various criteria.

Risk assessment (or risk comparison) follows the risk analysis and completes the risk assessment procedure.

In risk assessment, the resulting estimate of the risk level R is related to one (tolerance level R_{dop}) or several specified risk levels, which are determined on the basis of the tolerance level of risk. The risk tolerance level is determined by acceptable risk criteria.

If $R > R_{tolerance}$, the risk is considered unacceptable. The recommended typical risk levels (categories) and their ranges are shown in Table 1.6.

Table 1.6 - Typical risk levels (categories)

Level of risk	Value range
Unacceptable	$R >$ Rdop
Unwanted	0.1-Rdop $\leq R <$ Rdop
Allowable	0.01-Rdop $\leq R < 0.1$-Rdop
Not taken into account	$R < 0.01$-Rdop

The results of the risk assessment can be presented by means of a risk matrix, which is a table combining the frequency of an event and the severity of its consequences, to clearly inform decision-makers of the levels of risk for the event in question. The form (dimension) of the matrix depends on its application.

The risk matrix is constructed as follows:

– On the vertical axis, the probabilities (frequencies) of occurrence of the event are counted, presented as a scale (usually logarithmic) according to the adopted frequency levels (see Table 1.3);

– on the horizontal axis, the dimensions of the consequences of the event, presented as a scale (usually logarithmic) according to the adopted levels of consequence severity (see Table 1.4);

– The level of risk for each cell in the matrix is determined and ranked according to the risk tolerance criteria.

A typical form of risk matrix containing 6 levels of frequency and 4 levels of severity of consequences, in which the level of risk is ranked in 4 categories, is shown in Table 1.7.

Table 1.7 - Model form of risk matrix

Frequency levels	*Levels of risk*			
Frequent	*Unwanted*	*Unacceptable*	*Unacceptable*	*Unacceptable*
Probable	*Allowable*	*Unwanted*	*Unacceptable*	*Unacceptable*
Random	*Allowable*	*Unwanted*	*Unwanted*	*Unacceptable*
Rare	*Not taken into account*	*Allowable*	*Unwanted*	*Unwanted*
Extremely rare	*Not taken into account*	*Not accepted in consideration*	*Allowable*	*Allowable*
The unlikely	*Not taken into account*	*Not accepted in consideration*	*Not taken into account*	*Not taken into account*
	Minor	Irrelevant	Critical	Catastrophic
	Levels of consequence severity			

A more accurate representation of the results of the risk assessment, applied where necessary, can be a graph in logarithmic frequency-heaviness coordinates.

Based on the results of the risk assessment, a decision is made as to whether the risk should be treated and the priority for risk treatment. The decision to treat the risk is made by the management of the railway transport organisation on the basis of ethical, legal, financial and other considerations, including risk perception.

An example of the decisions made based on the results of the risk assessment for each level of risk is shown in Table 1.8.

Table 1.8 - Example of risk treatment decisions made

Level of risk	Solutions
Unacceptable	Risk must be eliminated. The treatment of risk is necessary.
Unwanted	The risk must be reduced. Risk treatment is necessary.
	The risk can be accepted with the agreement of the organisation's management, in cases where risk reduction is not feasible or not feasible. The treatment of the risk comes down to remediation.
Allowable	The risk is accepted with appropriate monitoring and control and with the agreement of the organisation's management. Treatment of the risk is not required or is limited to remediation.
Not taken into account	The risk is accepted without the consent of the organisation's management. No processing of the risk is required.

Uncertainty and sensitivity are central issues that arise when using the results of a risk assessment. Also, the importance of the contribution of the individual components to the overall risk assessment is important. Model uncertainty reflects the inherent weaknesses and inadequacies of the model and is a measure of its realism. Uncertainties in model inputs arise because of incomplete available data and the need to fill them in through expert judgement. When quantitative risk analysis requires an estimate of uncertainty in the final outcome, sensitivity studies are the simplest and most cost-effective approach.

The sensitivity s_j to parameter j is defined as the change in a quantitative measure of risk per unit change in that parameter, i.e:

$$S_j = \frac{\Delta R_j}{\Delta P_j}$$

where ΔR_j is the change in the risk measure resulting from a change in the jth parameter of the model;

ΔP_j is the change in the j-th parameter of the model.

For example, a 10% change in the internal locking failure rate (ΔP_j) could change the risk (ΔR_j) by a factor of 2. Then the sensitivity of the risk measure to the internal locking failure rate would be

$$S_j = \frac{2}{0,1} = 20$$

In theory, it is possible to test the sensitivity of a quantitative measure of risk to each of the parameters. But in practice for most quantitative risk analysis approaches this is not feasible due to the large number of parameters considered. Therefore, usually only sensitivities are determined for supposedly important parameters or for parameters that are known to have a high degree of uncertainty. The parameter of the model that has the greatest impact on risk will have the greatest sensitivity.

Identifying the importance of key inputs to overall risk is one of the most important tasks in the use of quantitative risk analysis.

The total risk (R) is the sum of the risks from all occurrences of adverse events (R_i):

$$R = \sum_{i=1}^{n} R_i$$

where n is the total number of adverse events, R_i is the risk from the occurrence of the i-th adverse event ($i = 1, 2, ..., n$)

For clarity, the components of the overall risk (e.g. the occurrence of undesirable events) are sorted in descending order of importance:

$[R_1, R_2, ..., R_n]$, where $R_i \geq R_{i+1}$.

The result is a list (tabulated presentation) of all undesirable events, ranked in descending order of contribution to the overall risk, clearly indicating the most important events for which risk reduction measures can be most effective.

1.3 Methodological approaches to risk assessment based on hazard, vulnerability and damage analysis

The general procedure of risk analysis for infrastructure and HRE involves a sequential analysis of the hazards to which the facility in question is exposed, an analysis of the vulnerabilities of the facility in relation to the hazards identified, and an analysis of the damage from the manifestation of hazards realized when the facility is found to be vulnerable [4, 5].

$$AR = AH \cup AV \cup AU.$$

Figure 1.8 shows a schematic diagram of a risk analysis of a railway infrastructure facility, using the following symbols:

– NS is the initial state of the object (system);

– s_0 is a scenario of an object (system) successfully performing its functions;

– KS_0 is the desired end state of the object (the end state when the object successfully performs its functions),

– ε_0 is the vicinity of the KS_0 point at which the end states can be considered acceptable (safe);

– IS_1, IS_2 - initiating hazardous events;

– PS_1, PS_2 - limit states of the object (system);

– s_i ($i = 1, 2, ..., N$) is *the* i-th failure scenario that occurs after one of the limit states is reached;

– KS_i ($i = 1, 2, ..., N$) is the unacceptable (dangerous) end state of the object (system) corresponding to scenario s_i;

– U(KSi) ($i = 1, 2, ..., N$) is the damage corresponding to the final state of KSi.

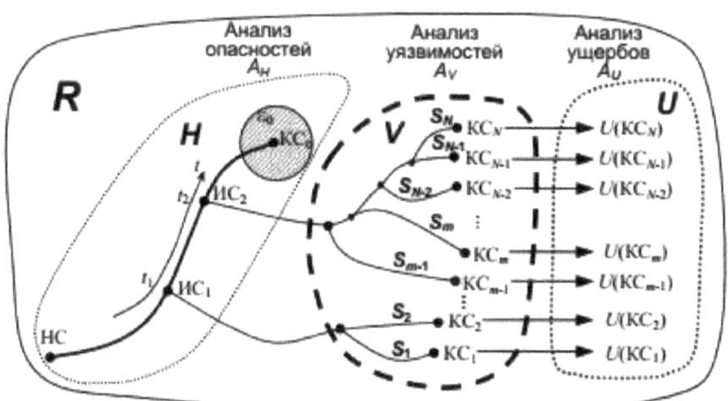

Figure 1.8 - Structure of risk and security analysis

The **hazards** for the infrastructure facility and HSE are understood as operational loads, element failures, external extreme (design and beyond design) impacts, operator errors, unauthorized impacts. Depending on the level of uncertainty and possible mechanisms of reaching limiting states of elements, hazards acting on the object are considered [6, 7, 8]:

– as random variables characterised by the probability that a hazardous event of a certain intensity will occur;

– as probability distributions that define the density distribution of the probability that hazardous events will occur in terms of intensity;

– as random processes to describe the history of loads and extreme impacts on an object.

The vulnerability of an object is characterized by a set of scenarios of random events (failures) and cause-effect relationships between these events, i.e. the structure of the scenario graph of the system [9, 10]. At that, the parameters of vulnerability of an object will be the conditional probabilities of realization of different final states of the object, arising in case of escalation of a failure, developing in the system after an initiating event of different type and intensity. Vulnerability analysis implies the study of event sequences and cause-effect relationships between events occurring after the initiating event until the object reaches its final states. In other words, vulnerability analysis consists of a qualitative and quantitative investigation of the structure of accident escalation scenarios. Thus, vulnerability analysis involves a detailed study of the scenario tree of the object in question.

The principles of constructing scenario trees describing accident escalation scenarios are studied in detail within the framework of scenario structuring theory. Among the approaches of scenario structuring theory, methods based on the construction of 'event tree' or 'influence diagram' type graphical models describing probabilistic causal relationships between events in the accident escalation process are central.

The trajectory s_0 in the state space describing the functioning of an object (system) from the initial state of the NS to the required final state of KS_0 is usually called a "success scenario" (Fig. 1.8). At time moments t_1, t_2, ..., t_k, initiation events IS_1, IS_2, ..., ISK may occur on the object, which are able to deviate the trajectory of the scenario from the curve s_0, triggering a sequence of events corresponding to failure scenarios s_1, s_2, ..., s_N, which will lead to the system reaching the corresponding final states KS_1, KS_2, ..., KS_N.

The implementation of a particular accident scenario s_i leads to the object reaching the corresponding unacceptable final state of KSi, associated with damage *U*(KSi). Thus, **damage** is the result of a change in the state of an object (system), expressed by one or more of the following categories

– total or partial loss of health or death of a person;

– damage to the integrity of an object or deterioration of its other properties;

– actual or potential economic or social losses resulting from any events, phenomena, actions;

– loss of property or other material, cultural, historical or natural values.

A distinction is made between direct, indirect, total, and total damage when considering damage from a facility accident [11, 12, 13].

The direct damage in an accident at a facility (in a system) is understood as losses and damages to the population, the natural environment and all structures of the national economy (including the system itself) which are caught in the zone of action of the striking and damaging factors of the accident. They are determined by the number of deaths and casualties among the personnel and population, irrecoverable losses of fixed assets, evaluated natural resources and losses, caused by these losses. When considering the structure of direct damage, direct economic damage, direct environmental damage and direct social damage are distinguished.

Indirect damage from an accident is the losses, losses and additional costs incurred by the population, natural environment and national economy objects, which were not in the hazardous zone of the accident at the facility, and caused by disturbances and changes in the existing structure of economic relations, infrastructure, as well as losses (additional costs) caused by the need to implement certain measures to eliminate the consequences of the accident.

Total damage is the sum of direct and indirect damage, as well as the costs of dealing with the consequences of the accident. The total damage is determined at a particular point in time and is intermediate in comparison to the total damage, which will be quantified in the long term.

In the following subsections the hazards, vulnerabilities and damage assessments of the railway infrastructure and rolling stock will be considered in sequence, using which the level of risk for the site in question can be assessed.

By successively assessing the hazards, vulnerabilities and damages for the railway infrastructure or rolling stock, the level of risk for the object in question can then be assessed

$$R = \vec{H} * V * U^T \quad 4.2.$$

where $\vec{H} = \{P[IS_1]; P[IS_2]; ....P[IS_K]\}$ - is a vector of hazards whose components are the probabilities of realisation of initiating events of IS_1, IS_2, ..., IS_K;

V = [P(KS_i |$ИC_j$)] - N x K vulnerability matrix, the components of which represent the probabilities of realisation of possible damaged states of the KSi, subject to various extreme impacts on the object by the IS_j ;

$U\{U(KS_1); U(KS_2); ....U(KS_N)\}$ - is a vector of damages whose components are the damage values corresponding to the end states of KC_1 , KC_2 , ..., KC_N .

1.4 Hazard analysis

Hazard analysis is usually the first stage of RRisk analysis of technical facilities [4, 8, 13]. Hazard for infrastructure facilities and HSE is a probabilistic characteristic that determines the possibility of impact on the object of damaging factors of a certain type, intensity and duration as a result of the implementation of some initiating event that can occur both at the facility itself and in the external environment.

It should be borne in mind that during the course of an accident at a facility, secondary hazards and the secondary damaging factors that they generate may arise, acting on the facility. The potential for the initiation of these secondary hazards will be determined by the vulnerability of the facility to the primary hazards. Therefore, hazard analysis should be performed in conjunction with an analysis of the vulnerability of the elements of the facility (system) to the hazards acting on them.

A hazard to infrastructure facilities and rolling stock is defined by a set of random events or processes Th: extreme external natural and manmade hazards, inappropriate actions of personnel and conditions of operation of technical systems of the facility that have the potential to lead to an accident. Examples of such events are: seismic activity or man-made disasters at nearby facilities (external hazards), depressurisation of a tank with toxic chemicals or accumulation of damage due to wear and tear (internal hazards).

A hazard is characterised by the effects of damaging factors on an object. It is a random variable which, in the simplest case, can be characterised by the probability of occurrence of a hazardous event P(Th) (Fig. 1.9(a)).

Where a more precise description of the hazard of an extreme event is needed, it should be characterized not by a point estimate of the probability *P(Th)*, but by the hazard intensity distribution curve p_{Th} (Ω) shown in Figure 1.9(b), or the integral distribution function p_{Th} (Ω) where Ω характеризует is the intensity of the occurrence of a hazardous event.

When deciding which physical parameter of the impact of a hazardous process on an object should be chosen to assess the intensity of the hazard, the vulnerability of the object to the action of the various components of the impact should be taken into account. For example, in the case of seismic effects on a site, one part of the site elements is most sensitive to ground vibration acceleration and another part to vibration amplitudes.

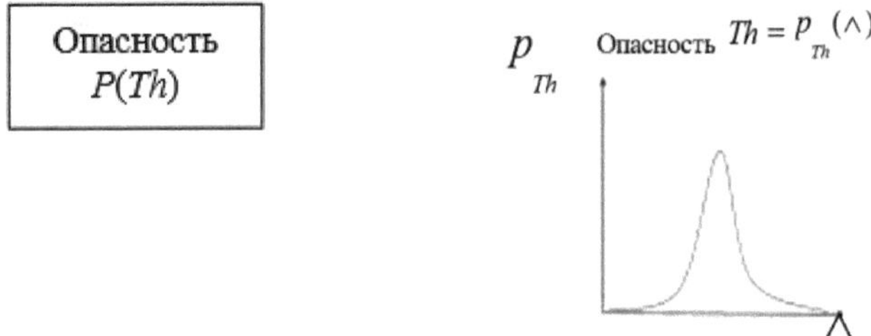

a) The probability of an elementary event is described by means of a point estimate

b) The probability of an elementary event is described by the probability density function of its intensities

Figure 1.9 - Description of probability of a hazardous event

Functioning is associated with storage and transportation of significant volumes of hazardous substances, transformation of significant volumes of energy, receipt of processing and transmission of powerful information flows, on-site hazardous technological processes, as well as external natural and man-made hazardous processes in the areas of the facility, which are expressed in extreme external impacts on the facility. A distinction is made between external and internal hazards depending on whether the source is outside or inside the boundaries of the infrastructure facility and HSE.

Internal hazards to infrastructure and rolling stock are initiated by hazardous processes whose potential is determined by the following main factors

– the mass and W composition of chemical, biological and radiological hazardous substances in the facility (stored or transported); - the amount of energy E circulating in the facility,

– The volume of incoming and outgoing information flows I.

Internal hazards of infrastructure facilities and HSE also include operational stresses on components, exposure to aggressive chemical environments, radiation, control system failures, etc. A significant segment of the spectrum of internal hazards is caused by human error (mistakes of personnel and decision makers, including violations of regulations, etc.).

External hazards include damaging factors, the effect of which is a consequence of dangerous natural and man-made events (processes) occurring outside the boundaries of the object. Such initiating events may be a seismic shock, a technogenic accident at a neighbouring facility, the effect of extreme weather conditions, etc. In addition, external hazards include events associated with disruptions to external energy, transport and other infrastructures that result in disruption of transportation and other technological processes, damage to on-site control and supply systems, as well as terrorist impacts on the facility.

The level of detail in the description of the hazards is determined, on the one hand, by their nature and, on the other hand, by the mechanisms involved in destroying the object. In the simplest case, the hazard can be described by means of a point estimate as the probability P_H that an object will be exposed to a damaging factor H with a given exposure intensity W_H. If a more precise description of the hazard is required, then a 'hazard curve' is used - a function $f_H(w)$ of the probability density distribution of the intensity of exposure to the H factor w (Figure 4.10).

It is often convenient to discretise a continuous 'hazard curve' by replacing it with a hazard vector whose components represent the probabilities of hazard H occurring at different levels of exposure intensity:

$$\vec{P}_H(w) = [P_H(w = W_1);\ P_H(w = W_2);\ ...;P_H(w = W_k)]$$

In both cases under consideration the hazards are understood as the damaging factors of extreme natural-technogenic-social phenomena, which are considered as random events characterized by the frequency of realization and intensity. In this description of a hazard, the time-duration factor of some hazardous process is replaced by a single peak extreme impact, which is considered to be decisive in terms of damage to the object. Examples are "hazard curves" for seismic effects, wind loads, hazardous substances, etc. These curves make it possible to estimate the probability for each level of intensity w of the impact factor H.

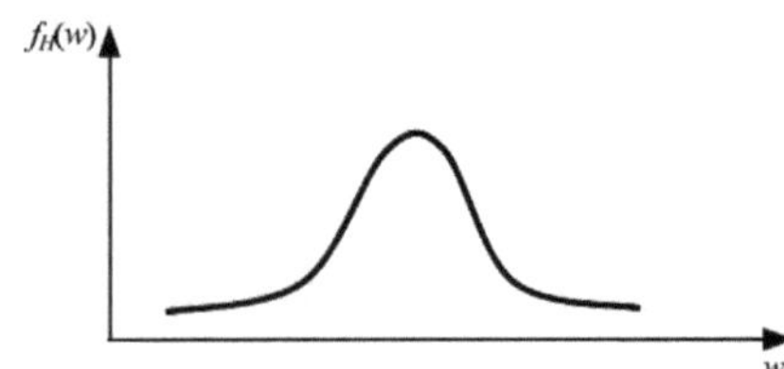

Figure 1.10 - Probability Distribution Density of the Intensity of Exposure to a Hazard

In this approach, the description of the hazard is limited to a one-dimensional distribution law of the probability that a random variable characterising the intensity of exposure to a damaging factor will exceed a given value over a given time period (usually one year). Knowing this law, it is possible to calculate more complex indicators (e.g. the probability of a random variable exceeding a certain critical level during a given period, interpreted as the lifetime of the infrastructure facility and HLW).

The occurring hazardous processes have probabilistic (random) nature and, as a rule, are non-stationary (Fig. 1.10). This is due to changes in the parameters of the object (system) at various stages of the life cycle, transient processes, as well as changes in technology, modernisation of the object, etc. The notion of mathematical expectation function m_H (t) and correlation function K_H (t, t^*) are introduced to characterise random processes. For any point in time $t=t^*$, the random process of an impact factor is a random variable characterised by mathematical expectation and variance.

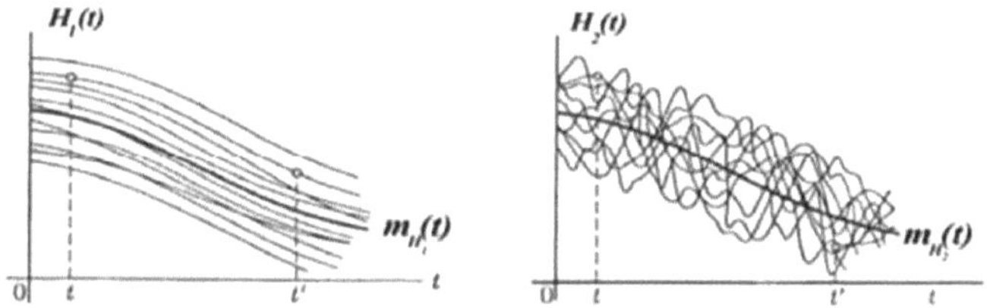

Figure 1.11 - Representation of hazards as random processes

Quantitative description of initiation, development of accidents at a facility can be carried out on the basis of fundamental laws of physics, chemistry and mechanics of catastrophes. In this case the stages of emergence and development of accidents can be characterized by different combination of physical, chemical and mechanical affecting factors.

In this formulation of the problem, the hazards for the infrastructure object and rolling stock will be characterised by a random vector, which is r a function of the vectors: internal and external force actions $Q(t)$, temperature actions Tr(t), fields of concentrations of hazardous substances cr(t), emissions r (t) and information flows Ir(t)

$$\vec{H}(t) = F\{\vec{Q}(t), \vec{T}(t), \vec{c}(t), \vec{\Psi}(t), \vec{I}(t)\}$$

Catastrophe mechanics, physics and chemistry, as fundamental scientific disciplines, should be considered as a scientific basis for analysis of sources of occurrence and scenarios of accident development. The data on the operating load conditions of the elements of facilities are initial in assigning the main design parameters included in the basic equations of catastrophe mechanics and subsequent modeling of initiation and development of accidents.

Quantitative description of the initiation and development of accidents at railway infrastructure and rolling stock objects can be carried out on the basis of fundamental laws of physics, chemistry and mechanics of catastrophes [4, 14]. At that, the stages of emergence and development of accidents can be characterized by various combinations of physical, chemical and mechanical affecting and damaging factors.

Analysis of most man-made and natural accidents in the world shows that they are caused by three main hazards:

– uncontrolled leakage of hazardous substances W;

– uncontrolled release of hazardous energy E (mechanical, thermal, electromagnetic, light);

– uncontrolled release or destruction of information flows I.

Taking into account the scale of accidents and disasters assessment used in international practice (local level - 1, facility level - 2, local level - 3, regional level - 4, national (federal) level - 5, global level (transboundary) - 6, planetary level - 7) as well as the above parameters *W, E, I,* we can construct limiting areas of their hazardous states to categorize infrastructure facilities and rolling stock. At that, a quantitative indicator of object hazard will be a modulus of the radius-vector in the space *W, E, I*

$$|B_H| = \sqrt{\vec{W}^2 + \vec{E}^2 + \vec{I}^2}$$

where $\vec{W}$, $\vec{E}$, $\vec{I}$ - the hazards of the facility discussed above, expressed on a 7-point scale of catastrophe classes. In general, the numerical value of this hazard will vary from 1.73 to 12.2.

In the traditional formulation of the problem, hazard analysis for the infrastructure and rolling stock is the first step in solving the problem of assessing the risks associated with accidents at these facilities. But it is also of interest to solve the inverse problem, in which hazards to a facility are classified based on the known consequences of accidents that have occurred at the facility.

The intensity of hazards is divided into the following groups when dealing with such tasks:

- Group O1: hazards causing hypothetical accidents, which may occur under unpredictable scenarios with maximum possible damage (total destruction of the facility) and a high number of casualties;

– group O2: hazards causing design-basis accidents involving irreversible damage to important site elements with high damage and loss of life;

– group O3: hazards causing design basis accidents, accompanied by out-of-range conditions with foreseeable and acceptable consequences;

– Group O4: hazards which cause a facility accident and which cause deviations from normal operating conditions during normal operation of the facility;

– group O5: hazards in which the facility operates normally.

Table 1.9 presents the hazard characteristics for the facility according to the types of accidents.

Table 1.9. -Types of accidents and hazard characteristics

Type of accident	**Extent to which the hazards are known**	**Group hazards**
Hypothetical	A combination of unknown low-probability and hard-to-predict design, technological triggering events and striking factors of enormous intensity, including terrorist impacts.	O1
Project	The impacting factors and initiating events and the development of the injuries are not fully known.	O2
Project	The factors involved are known and foreseeable.	O3
Regime (deviations from normal conditions)	The contaminants are studied and controlled.	O4
No (normal (standard) operating conditions)	The attacking factors are well understood and manageable.	O5

Hazard analysis for infrastructure facilities and HSE should be based on taking into account, assessing and analyzing (predicting) all possible accidents leading to losses at the facility throughout its life cycle. Therefore, the accuracy of the analysis results is determined by the completeness and accuracy of the technical and technological description of the facilities.

For the purpose of hazard analysis, information on the composition, purpose and technology of the operating equipment should be obtained at each stage of the facility's life cycle. In carrying out the hazard analysis procedure, a database including information should be collected for the facility under construction:

– The geographical location of the facility; maps of natural and man-made hazards in the area where the facility is located, defining the topographic, natural, man-made, and climatic conditions at the location of the facility. The above information defines the "external environment" parameters for the facility and is used both to define input parameters in hazard analysis procedures (e.g., wind direction and strength, temperature conditions, etc.) and to justify the exclusion of unimportant factors from consideration in hazard analysis procedures;

– the characteristics of the technological (transport) process. This information is used to select possible models of the accident phenomena and to determine possible damaging factors;

– The composition of the installed process equipment, describing the details of the processes associated with each installation. This information is used to identify the hazards associated with accidents in that plant;

– the location of the equipment on the site (master plan of the facility) indicating coordinates, mutual distances, dimensions. This information is used to determine the possible impact of accident hazards on neighbouring process units and to determine the parameters of a possible accident escalation;

– the reliability and failure rates of the equipment provided by the supplier. This information is used to determine the expected failure rate of the piece of equipment in question;

– Accidents, incidents and their consequences, both at the facility and at other similar facilities. This information is used to determine the expected frequency of accidents at the facility;

– The planned and available accident prevention and response facilities. This information is used both to determine the expected frequency of accidents at the facility and to obtain the necessary inputs to the risk analysis procedure for response time and other necessary inputs.

The following requirements are imposed at various stages of the life cycle:

– At the stage of justification of the decision to construct the facility, the basic process flow diagram used at the facility, the expected volumes and composition of hazardous products and raw materials to be transported at the facility, and the typical composition of the equipment must be known;

– At the detailed design stage, a complete process flow diagram and a complete list of installed equipment with known characteristics and technical data sheet provided by the manufacturer must be known;

– During the operational phase, documented information on pre-accident and emergency situations, equipment failures, human errors, possible process changes and replacement of worn-out equipment, the actual volumes and parameters of hazardous substances handled at the facility should be accumulated.

These requirements necessitate the use of the following sources of information at various stages of the life cycle of a railway facility:

– at the stage of justification of the decision on construction of the facility - data of the world statistics of failures and accidents at the facilities of similar purpose. The quantitative physical quantities required for the hazard analysis are calculated on the basis of a known basic process flow diagram of the processes;

– at the stage of detailed design, construction and commissioning - the same as at the stage of decision justification for the construction of the facility, taking into account the manufacturer's technical data sheets for equipment on reliability, average MTBF. Quantitative physical quantities required for hazard analysis are calculated on the basis of a complete process flowchart;

– In the operational phase, the same as in the decision-making and engineering design phases, taking into account the actual equipment failure statistics identified in the operational phase. At this stage, the statistical contribution of personnel errors to the value of real reliability of technological processes is also identified. Quantitative physical values required for hazard analysis are calculated on the basis of the real technological scheme of processes, real reliability parameters.

The main causes and factors contributing to equipment failure hazards include:

– the hazards inherent in the processes taking place at the site;

– physical wear and tear, corrosion, mechanical damage, temperature deformation of the equipment;

– The supply of energy resources (electricity) is interrupted;

– external man-made, anthropogenic and natural influences.

For the purposes of hazard analysis it is reasonable to divide the whole variety of possible accident causes into a limited set of standardized accident initiation models, characterized by deterministic physical parameters (diameter of the equivalent orifice, type of flow, type of product, etc.) and probabilistic parameters (conditional probability and frequency of realization of a given event). It must be kept in mind that in this procedure (discretization procedure) the accuracy of the description of the physical parameters of the accident initiation models may be lost.

All equipment used on infrastructure and rolling stock can be divided into a limited number of categories, according to its physical and chemical processes and design features. Within one category, equipment is characterized by the same set of possible accident initiation patterns. For each category of process equipment, specific initiation events and models of destruction of equipment under the influence of accidental hazards (impact and thermal effects, shrapnel damage, etc.) should be developed.

During the decision-making, working design and operational phases of facility construction, the quantification of triggering event models and their expected frequencies is refined on the basis of more complete information:

- about the design or actual parameters of the technology;
- about perceived or actual reliability parameters;
- on the safety management system proposed or in use;
- on the planned or actual level of staff action.

1.5 Vulnerability analysis

The concept of vulnerability is now increasingly used in risk assessment of technical systems in order to characterise the response of the systems in question to extreme impacts. However, there is no single established definition of vulnerability in risk theory. Generally, vulnerability is understood as the openness of a system to various extreme internal and external events/influences that contribute to a catastrophic process. Quite often the notion of vulnerability is defined through the characteristics of the system associated with it. For example, the vulnerability of a system is understood as a set of properties that are the opposite of system stability and survivability, as well as its ability to perform specified functions in the event of partial damage.

Since the concept of vulnerability is multifaceted and must reflect the physical, organisational, technological and functional aspects of the state of the system, it is highly relevant to develop a unified methodological approach to the quantification of vulnerability, i.e. to determine the measure of vulnerability for technical systems.

Changes occurring in a technical system designed to produce a certain result (or to implement a given technological process) can be represented as a trajectory in the system state space Ω defining the transition from the initial state of the NS system to its final state KS0 (Figure 1.12. a)). In cases where such a transition can be achieved, the system is said to have a given scenario (or "success scenario") S0 [4, 9, 10]. The final state of KS0 defines a set of values x_0^1; x_0^2; L, x_0^m; that the system state variables x_1 , x_2 , L, x_m must take in order for the system to meet the requirements imposed on it. These requirements may, for example, include: the structural integrity of the system, the intactness of its elements, the system performing specified functions, ensuring specified performance and quality (of products or services), etc. The listed system state variables determine the dimensionality and configuration of the system state space.

If some IS triggering event$_1$ occurs in the system, it can deviate from the scenario S_0 (Fig. 1.12 b)) and go on to implement a new scenario S_1 , ending in a KS final state$_1$, different from the given (desired) KS final state$_0$:

$$KC_1(x_1^1, x_1^2, ..., x_1^m) \neq KC_0(x_0^1, x_0^2, ..., x_0^m)$$

In this case, the system can be said to have failed and is unable to provide the required KS_0 end state. That is, the system has become vulnerable to an IC_1 triggering event. Due to the high level of uncertainty regarding the type and intensity of initiating events and the ability of the system to resist initiating events, the measure of vulnerability must be probabilistic, i.e. determined by the failure probability (O) of the system:

$$Vf(P[O]).$$

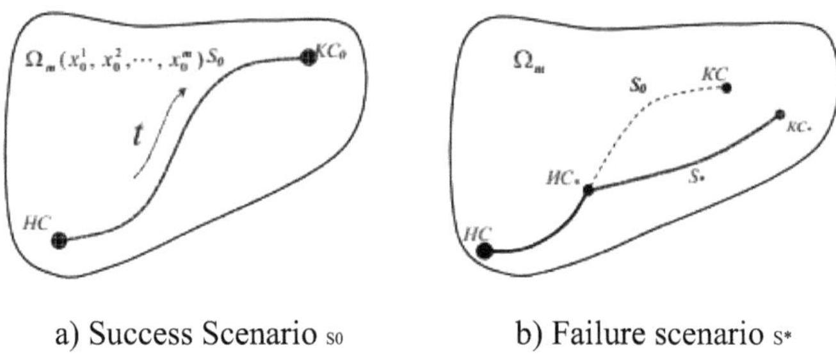

a) Success Scenario S_0 b) Failure scenario S_*

Figure 1.12 - Success and failure scenarios

Since the vulnerability properties of a system do not become apparent until after some abnormal triggering event has occurred in it (or the system has been subjected to some abnormal impact), the measure of vulnerability must be determined by the conditional probability of failure of the system, assuming the system has been subjected to a triggering event $Vf(P[O \mid IS]$.

Obviously, in practice it is not always possible to ensure that the system reaches the given final state of KS_0 absolutely precisely. Real systems will always be subject to some, sometimes weak, initiating influences that will slightly deviate the scenario trajectory from the given s_0 success scenario. In addition, the deviation from the s_0 success scenario will also be due to natural variation in system parameters. Therefore, the vulnerability assessment must deal with the conditional probability of the final state of the system falling outside the given region ε_0 of the state space Ω_m. In particular, Fig. 1.13 shows a weak initiating event ICk which leads to a final state CCk lying within the region ε_0. It can be assumed that the system is not vulnerable to the initiating event of ICk.

Further, system failure is defined as the exit of the final state of the system from the area ε_0 of the system state space Ω_m, outside which the system does not provide the specified functions or quality of products (services).

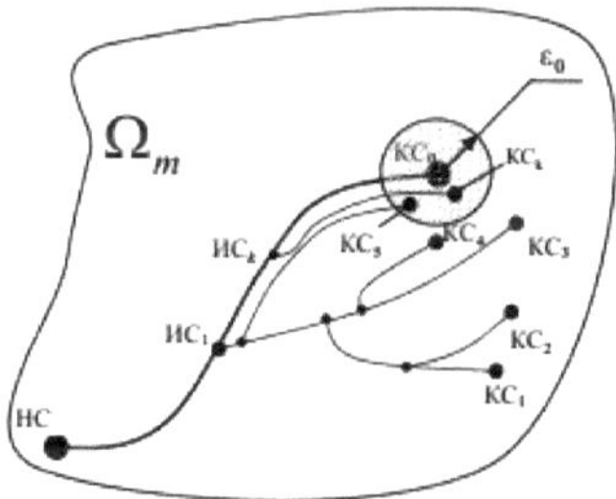

Figure 1.13 - Area of conditionally intact states ε_0

On the basis of the above, the following definition can be formulated: System vulnerability is the conditional probability that the final state of the KS* system will leave the boundaries of the given area ε_0 of the state space Ω_m of the system in the event that a triggering event of the IS* occurs

$$V = P[(||KC_* - KC_0|| > \varepsilon_0)|ИС_*]$$

The choice of the state space of the system and the way of defining the metric in this space. The Euclidean metric of the space Ω_m can be chosen:

$$\|KC_* - KC_0\| = \sqrt{(x_*^1 - x_0^1)^2 + (x_*^2 - x_0^2)^2 + \cdots (x_*^m - x_0^m)^2}$$

or

$$\|KC_* - KC_0\| = \max\left\{(x_*^1 - x_0^1)^{\square}; (x_*^2 - x_0^2)^{\square}; \ldots (x_*^m - x_0^m)^{\square}\right\}$$

In particular, different one-dimensional system state spaces Ω1 can be chosen to describe different aspects of the system state: physical, functional and economic. Therefore, the physical, functional and economic components of vulnerability can be distinguished, which can be defined in the corresponding one-dimensional state-spaces of the system.

If a one-dimensional space $\Omega 1^{ph}$ is chosen as the state space of a system whose single coordinate determines the physical degree of damage to the system when the system reaches the final state KS* ($x^1 = DS_*$), then the physical vulnerability V_{ph} of the system is the conditional probability that the system receives a certain degree of damage (probability of the final state that corresponds to this degree of damage)

$$Vph=P[DS> DS_d \mid IS_*]$$

where DS_d is the allowable degree of damage to the system.

If one-dimensional Ω_1^U is chosen as the state space of the system , whose single coordinate determines the level of damage that occurs when the system reaches the final state of KS_* (x^l =U), then the economic vulnerability of the system V_e is defined as the conditional probability that in time Δt a triggering event occurs in the system resulting in damage that exceeds a given threshold value U_d

$$V_B = P[U>U_d \mid IS^*]$$

If a one-dimensional space $\Omega_1^{\Phi_*}$ is chosen as the space of system states, the only coordinate of which determines the extent to which the system performs the given functions that are implemented by the system when it reaches the final state of the CS* (x_1 = F*), then the functional vulnerability V_F is the conditional probability that the system loses its ability to perform certain functions if the in Ω_1 initiating influence

$$V_{\Phi} = P[(||\Phi_* - \Phi_0|| > \delta)|ИС_*]$$

where F_0 is the given functional level, δ is the permissible deviation.

Thus, the values V_{ph}, V_e and V_F can be regarded as physical, economic and functional components of the system vulnerability vector, and the assessment of system vulnerability involves determining the conditional probability of failure in the system, assuming that a triggering event has occurred.

The vulnerability of a system is often defined as the opposite (in a probabilistic sense - addition to 1) of robustness and resilience [15].

System robustness refers to the conditional probability that, in the case of an initiating action by IS^*, the final state of the system KS^* deviates in the state space from KS_0 by an amount not exceeding a given small value ε_0

$$Rob = P[(||KC_* - KC_0|| < \varepsilon_0)|ИС_*]$$

It is then possible to arrive at the above definition of species vulnerability through the notion of robustness:

In dynamical systems, we should rely on the notion of stability (resilience), which describes the ability of a system to adapt and find a new stable position close enough to the target position so that it can perform the specified functions, following a perturbation.

In the probabilistic formulation, the recovery capability (adaptivity) of the system is understood as the conditional probability that the current state of the system $F(t)$ within a given time interval Δt after the initiating action should come to some steady state $F_c(t + \Delta t)$, close to the target state (Fig. 1.14)

$$R_S = P[(F(t) \to F_C(t + \nabla t)) \cap (F(t) \epsilon \delta)]$$

where δ is the region of stable (admissible) states.

Vulnerability can then be defined as the addition of system adaptability to unity:

$$V = 1 - R_{\downarrow}S = 1 - P[(F(t) \to F_{\downarrow}C\ (t + \Delta t)) \cap (F(t) \epsilon \delta)$$

Among the existing interpretations of vulnerability is the notion of structural vulnerability, which refers to the conditional probability that a network structure will not be able to perform its functions (i.e. the OS structure will fail) in the event that individual elements of the structure are disabled (element failure - SE).

$$V_{Str}\ \mathrm{P[OS \mid SE]}\ .$$

In the failure tree approaches, the vulnerability of a system is defined by the set of minimum accidental combinations that lead to different end states of the system, which differ from the given state of the CS_0 . Here, the minimum fault combination is the smallest set of initial events at which a faulty final state of the system is reached.

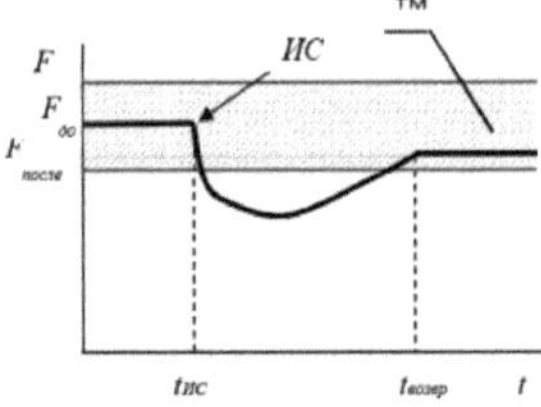

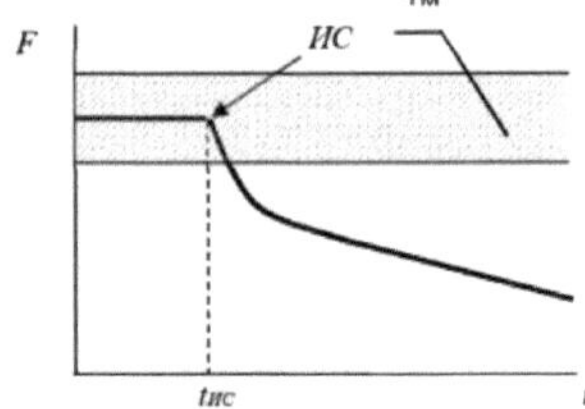

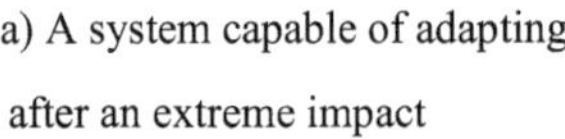

a) A system capable of adapting after an extreme impact

b) A system unable to adapt after an extreme impact

Figure 1.14 - Vulnerability of adaptive and non-adaptive systems

The complete set of minimum crash combinations in the tree represents all the combinations of events in which an accident could occur. The minimum trajectory is the smallest group of events at the occurrence of which an accident occurs.

Considering that the intensity of the triggering event *h* is not known in advance and may vary, the vulnerability of the system can be characterised by a conditional failure probability function (which is defined as the exit of the final state of the system outside the tolerance area ε_0), provided that the system is subjected to a triggering event of intensity *h* (Figure 1.15)

$V^I(h) = P[||KC_* - KC_0|| > \varepsilon_0|h]$

Given the definition of failure adopted above, this expression can be written in a shorter form:

$$V_I(h)\ PO\ \eta$$

The definition of Type *V* vulnerability$_I$, which takes into account the uncertainty associated with exposure intensity, is called a Level I vulnerability definition.

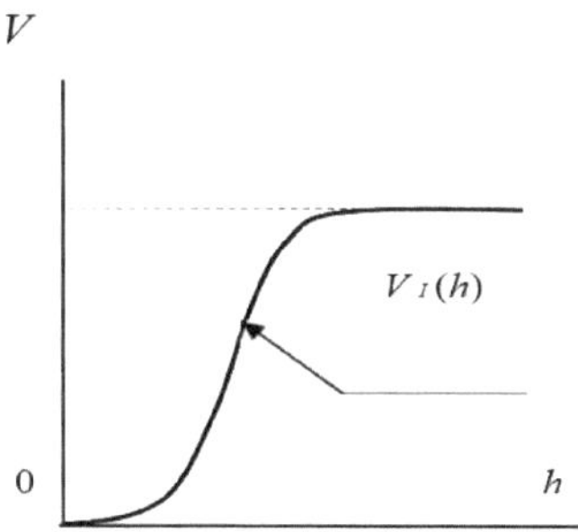

Figure 1.15 - System vulnerability curve

The system vulnerability curve shown in Figure 1.15 is the relationship between the probability of failure and the intensity of the initiating action.

If *N* end states are allocated to a system, the vulnerability of the system will be described by a family of vulnerability curves.

It should be noted, that if different failure scenarios can be realized in the system, developing after different initiating influences, and leading to different final states of the system, then the approach based on definition of the vulnerability of a species does not allow describing all the variety of options of non-performance by the system of its functions. In this case, the vulnerability of the system is not reduced to the above listed indices or individual characteristics of the system's openness to extreme influences, but is characterized by the structure of the tree of failure scenarios (Fig. 1.16).

In other words, system vulnerability is characterised by a set of scenarios of random events (failures in the system) and the causal relationships between these events. The vulnerability of a system is determined by the probabilities of realisation of the various end states of the system arising in the event of an escalating accident developing in the system after an initiating event of various types and intensities.

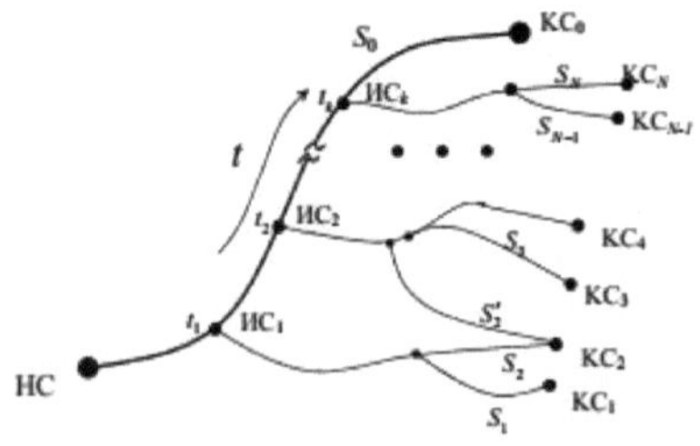

Figure 1.16 - Failure scenario tree

Vulnerability analysis involves the investigation of event sequences and cause-effect relationships between events occurring after the initiating event until the system reaches its final states. In other words, system vulnerability analysis consists of a qualitative and quantitative study of the structure of accident escalation scenarios. The principles of scenario trees describing accident escalation scenarios are studied in detail in *the* framework of *scenario structuring theory*, which will be presented below. Thus vulnerability analysis involves a detailed study of the scenario tree of the system in question.

Among scenario structuring theory approaches, methods based on graphical models such as event trees or influence diagrams that describe probabilistic causal relationships between events in the accident escalation process are central.

Consider some technical system that is designed to ensure the performance of a given function (achievement of a certain final state). The trajectory in the state space describing the evolution of the system from the initial state of NS to the required final state of KS_0 will be called a success scenario s_0 (Fig. 1.16). At time moments t_1, t_2, ..., t_k in the system, triggering events IS_1, IS_2, ..., ISK may occur which are able to deflect the trajectory of the scenario from s_0, thereby triggering a sequence of events corresponding to failure scenarios s_1, s_2, ... s_N, which will lead to the system reaching the corresponding final states KS_1, KS_2, ... KS_N. It should be kept in mind that different failure scenarios can lead to the same final state of the system (e.g., in Fig. 1.16 scenarios s_2 and S2'приводят to the final state KS_2). The vulnerability of the system can then be represented as matrix **V**, the components of which $v_{i,j}$ will represent the conditional probabilities of the system reaching the final state of KS_i, provided that the initiating event IS_j has occurred ($V_{i,j}$ P[KK_i | IS $_j$]):

By constructing a vulnerability matrix, the level of risk to the system in question can then be assessed.

It should be noted that an important task for assessing and reducing the vulnerability of railway infrastructure facilities and rolling stock is to identify the most critical, so-called "disproportionate" scenarios, which are characterized by achieving the maximum degree of damage to the system as a result of its small local damage. The solution of this problem requires creation of an algorithm for searching critical failure scenarios by constructing hierarchical models of the system structure and revealing the hierarchy for identification of critical (disproportionate) failure scenarios. This problem is solved, for example, in the framework of the theory of structural vulnerability developed in recent years, which focuses on the analysis and prevention of collapse scenarios of complex objects [16].

CHAPTER 2. DEVELOPING A MATHEMATICAL MODEL FOR RISK ASSESSMENT USING RISK MATRICES

2.1 Damage analysis

U damage from dangerous situations at railway infrastructure and rolling stock of man-made, natural and terrorist nature is determined by three basic components:

$$U\ UT\ US\ U\ N,$$

where UT is damage to technosphere objects;

US - damage to the environment;

UN - damage to the population (the individual and society as a whole).

UT damage is determined by summing up damages from damage and destruction of rolling stock and HSE infrastructure, industrial buildings and structures UTP, damages from damage and destruction of civil (residential) facilities UTG, damages from damage and destruction of external HSE facilities (transport, energy, pipeline and other systems) UTI:

$$UT = UTN + UTG + UTI.$$

US damage to the environment is determined by summing up the damage caused to the soil USp, the aquatic environment USa, the air environment USc, the flora USr and the fauna USg:

$$US = USn + USa + USv + USr + USj.$$

The damage to the UN population consists of the loss of human life and damage to health caused by UNHCR.

$$UN = UNj + UNz.$$

Injuries are divided into two groups according to the nature of their occurrence:

- direct (primary), associated with direct impacts of U_1 hazards at railway transport facilities;

- indirect (secondary), associated with the consequent in time manifestation of U_2 hazards at railway transport facilities.

Thus, damage can be seen as the sum of direct and indirect damages

$$U = U_1 + U_2.$$

The basic damage components can be defined separately for the primary and secondary impacts of the hazards:

- for U_T: $U_{TN} = U_{TNP1} + U_{TNP2}$,

$$U_{TG} = U_{TG1} + U_{TG2},$$

$$U_{TI} = U_{TI1} + U_{TI2};$$

- For U_S: $U_{Sn} = U_{Sn1} + U_{Sn2}$,

$$U_{Sa} = U_{Sa1} + U_{Sa2},$$

$$U_{Sc} = U_{Sc1} + U_{Sc2},$$

$$U_{Sr} = U_{Sr1} + U_{Sr2},$$

$$U_{Sj} = U_{Sj1} + U_{Sj2};$$

- for U_N: $U_{Nz} = U_{Nz1} + U_{Nz2}$.

Damage is quantified by two types of parameters:

– natural units - scales (number of damaged facilities and affected people, area of contaminated and damaged territories);

– equivalent economic units (roubles).

Direct (primary) and indirect (secondary) damages, taking into account the basic components, are determined for the following calculation cases:

– for the time of occurrence and development of actual hazardous situations UF; - for the time after realised emergencies *UE*;

– for the predicted time of possible emergencies *UB*,

$$U = U_F + U_N + U_V.$$

In deterministic assessments of primary and secondary damages, the feasibility studies for projects and production plants, technical and regulatory documents and cost estimates are taken into account.

Statistical estimates of primary and secondary damage use generalised information on emergencies contained in state reports from the Ministry of Emergency Situations, the Ministry of Transport, Gostekhnadzor, and the State Industrial Safety Service of Uzbekistan, as well as information from industries and agencies.

Probabilistic assessments of primary and secondary damage use simulation modelling data, data on the likely zones of action of the damaging factors, probabilistic statistical data on the vulnerability of facilities, the environment and the population in various hazardous situations.

Data extrapolation methods as well as expert judgement methods can be used to predict primary and secondary damages.

Detailed assessments of secondary damage should reflect trends towards a phased transition to the use of insurance mechanisms, the introduction of measures to prevent and monitor hazards.

The implementation of a particular accident scenario $s_i (i = 1, 2, ..., N)$ leads to the object (system) reaching the corresponding damaged final state KSi, associated with damage U(KSi) (see Fig.2.1).

Damage assessment consists of determining the value of the damage in kind or in money (economic valuation). The main types of consequence are:

- loss of life or harm to public health;

- socio-economic consequences (loss of a particular type of property, costs of resettlement, compensation for victims, lost profits from unconcluded and terminated contracts, disruption of normal economic activity, deterioration of people's livelihoods);

The main components of direct damage (Figure 2.1):

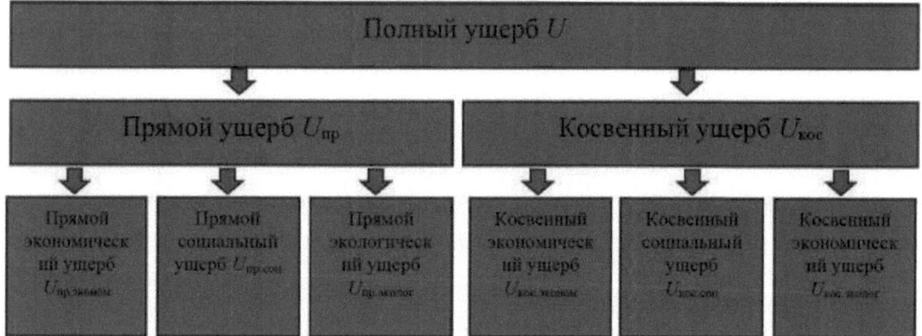

Figure 2.1. - Structure of total damage

Direct damage resulting from an accident at a facility is usually understood as losses and damages to all structures of the national economy that are affected by the accident. When considering the structure of direct damage, direct economic damage, direct environmental damage and direct social damage are distinguished.

The main components of indirect damage (Figure 2.1): Indirect damage includes losses incurred outside the direct impact area of an accident. Like direct damage, indirect damage is divided into economic, environmental and social damage.

Economic indirect damage includes the following components:

– Change in the volume and structure of industrial output (by type);

– changes in performance indicators in industry;

– premature retirement of fixed production assets and production facilities;

– the damage caused by the forced reconfiguration of control systems (additional costs of using spare control stations, additional costs of using mobile communication equipment).

The damage structure proposed in (Fig. 2.1.) makes it possible to represent the multicomponent damages from a real or hypothetical accident at a railway transport facility as a vector of damages, which will correspond to a certain end state U(KSi) of the "infrastructure object or rolling stock - environment" system

When organizing activities aimed at ensuring the protection of infrastructure and rolling stock, it is necessary to assume that the operation of such complex technical systems without failures is fundamentally impossible. When carrying out activities aimed at improving site security, measures should be developed not only to reduce the probability of incidents and accidents, but also to minimize their consequences [11].

Because of the high level of uncertainty in adverse scenarios, probabilistic damage estimates are preferable. The practically unlimited number of accident escalation scenarios and possible final states of the system requires the division of scenarios into major groups and their ordering by the level of expected damage, which makes it possible to construct a damage distribution curve.

The procedure for drawing the damage distribution curve is as follows.

a) Failure scenarios are identified s_i, $i = 0, 1, 2, ..., N$

of the system (scenario s_0 corresponds to the system successfully performing its functions with zero damage). For each scenario s_i, the probability of its occurrence p_i and the size of the consequences u_i are determined.

b) The table of received data is ordered by increasing severity of consequences $u_1 \leq u_2 \leq \ldots \leq u_N$ and an additional column is entered in which the probability of the damage exceeding the value u_i is entered:

$$P_i(U > u_i)\begin{cases}\sum_{l=i+1}^{N} p_l = P_{i+1} + p_{i+1}, i < N; \\ 0; i = N\end{cases}$$

This produces a table of N+1 rows (from 0 to N) and 4 columns: scenario, scenario probability, damage and the probability that the damage (if any of the scenarios is realised) may exceed the damage of a given scenario (Table 2.1).

Table 2.1 - Ordered list of scenarios

Script	**A measure of opportunity**	**Consequence s**	**Probability of exceeding the value *ui* : *P(U ui)***
S0	p0	ul	P0 1
S1	pl	ul	P1 P2 pl
S2	p2	u2	P2 P3 p2
...	...	...	...
Si	*Pi*	*Ui*	*Pi Pil pi*
...	...	...	...
*SN*1	*pN*1	*uN*1	*PN*1 *PN PN*1
SN	*pN*	*uN*	*PN pN*

In this case, each scenario s_i can be considered as a point u_i; Ріна of the plane "damage - probability of exceeding damage" (u, 0, P) . Then, by plotting a set of points on the plane (u, 0, P) u_i; P_i i 0, 1, 2, L, N, a stepwise damage distribution function can be obtained (Figure 2.2).

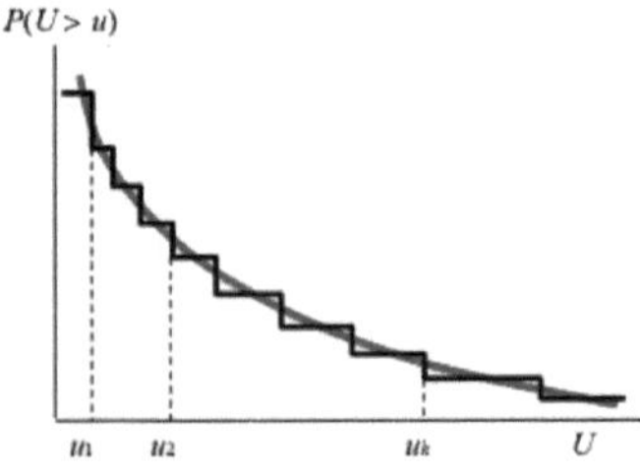

Figure 2.2 - Distribution of damages according to accident scenarios at the facility

c) Next, the envelope of the step diagram is constructed, which is an integral characteristic of the damages for the object in question.

Obviously, the shape of the distribution curve $E[P(U > u_i)]$ itself will have some level of uncertainty both about the values of the probabilities $P(U > u_i)$ of exceeding a certain damage value, and about the values of damages u_i corresponding to a certain exceedance probability (Figure 2.3).

Figure 2.3 - Uncertainty in the probability and magnitude of damage

Damage reduction activities are limited to ensuring that the damage distribution curve shifts to the origin of the coordinates. Figure 2.4. shows damage distribution curves for the same object under two protection strategies:

- The operating strategy for $£_{min}$ includes a minimum set of protective measures;

- The $£_{max}$ operating strategy provides the most comprehensive set of protective measures.

Obviously, the mathematical expectation of damages for the $£_{max}$ strategy will be significantly lower than for the $£_{min}$ strategy.

Figure 2.4 - Displacement of the damage distribution curve when implementing protection measures

It is common practice to distinguish between damage from routine, design basis and beyond design basis/hypothetical accidents at a facility when selecting measures to reduce expected damage:

- Regulatory accidents involve deviations from normal operating conditions during normal operation of a facility and result in minor damage;

- design basis accidents, which are characterised by out-of-normal modes of operation of individual units of a facility, accompanied by predictable and acceptable damage;

- for design-basis and hypothetical accidents involving irreversible damage to critical facility elements with high damage and loss of life. And some scenarios of such accidents can lead to complete destruction of the facility and escalation of the accident beyond the facility, accompanied by enormous indirect damage to the area where the facility is located and to the country as a whole.

An appropriate set of protective actions can be formed for these types of accidents:

- Measures aimed at reducing damage from regime accidents should be focused on improving technological processes, increasing the reliability of facility elements and the quality of their technical condition monitoring and repair work, as well as improving the qualifications of personnel.

- measures to reduce damage from design basis accidents, in addition to those listed above, also involve improving the survivability of the facility (including improving the system topology, introducing redundancy, improving control systems and establishing protection systems).

- measures to reduce damage from design-basis and hypothetical accidents, in addition to those listed above, include rational siting of the facility, taking into account the location of other facilities in the potential zone of impact from an accident at the facility, strengthening forces and improving emergency response procedures, increasing the availability of emergency services, establishing reserve funds, insurance and reinsurance mechanisms.

Measures to minimize the consequences of accidents should be classified according to their timing:

- Measures taken during the design, construction and operational phase (prior to an accident): rational siting of the facility; establishment of monitoring systems and comprehensive protection systems; organisation of continuous monitoring of critical elements and periodic inspections of their condition; organisation of preventive repair and replacement procedures for damaged elements, following routine inspections; identification of the most dangerous accident scenarios and construction of special protections designed to minimise the probability of realisation

- measures taken during the escalation of an accident: activation of protection systems, emergency shutdown of the facility, implementation of an emergency plan;

- post-accident measures: rescue work, emergency response measures, recovery work.

Measures to minimize damage at the facility should be divided into actions aimed at reducing economic, social and environmental damage. Reduction of economic damages implies efforts to protect fixed assets from the effects of potential accidental hazards and to increase the resilience of technological processes to failures. Social damage reduction is achieved through the protection of the site personnel and the population of the adjacent territories, rational siting of the site and the creation of sanitary zones around the site. The reduction of environmental damage includes efforts to protect water bodies, atmosphere, soil cover from ingress of hazardous substances and energies released during the escalation of possible accidents.

Practical measures to reduce damage from potential accidents at HSE facilities are based on specific preventive measures of scientific, engineering and technological nature implemented to counteract natural and man-made hazards. **A** significant part of these activities is carried out within the framework of engineering, radiation, chemical and medical protection of the population and territories adjacent to the sites of the facility.

Damage mitigation measures are site-specific.

However, these measures have common scientific, engineering and technological foundations that serve as a methodological basis for damage reduction. As examples of possible measures, the following can be named: improvement of technological processes, increase of reliability of technological equipment and operational reliability, timely renewal of fixed assets, use of high-quality design and technological documentation, use of high-quality raw materials, materials, components, involvement of highly qualified personnel, creation and use of effective systems of technological control and technical diagnostics, accident-free stopping of the production process, and improvement of the quality of the production process.

One area of effective damage reduction is the construction and use of containment structures.

Another direction of measures to reduce damage from accidents are measures to improve the physical resilience of facilities. In particular, seismic construction in earthquake-prone areas should be noted, as well as seismic reinforcement of buildings and structures constructed earlier without regard to seismicity. Earthquake-proof construction is carried out based on the maximum strength of earthquakes possible in the areas of construction. This area of preventive measures also includes measures to improve the physical resilience of facilities.

A distinctive feature of rail transport facilities is their integration into the interconnected transport, energy and telecommunications infrastructures that support the livelihoods of the population and economy. Accidents at HSE facilities that are part of the relevant infrastructure network transport systems can lead to cascading failures when these systems are operated under high load conditions. A separate block of measures should therefore be dedicated to reducing cascading damage associated with disruption to network infrastructures. The block should include a set of measures:

- to improve the topology of network infrastructures with the introduction of additional links allowing load redistribution in case of failure of individual elements;

- by increasing the marginal load reserve, allowing the system elements to withstand the additional loads resulting from load redistribution following the failure of individual elements;

- To develop a system of autonomous sources of supply of basic services and resources to the population and facilities of the economy necessary for their livelihood; and to establish emergency reserves in case of major accidents and related interruptions in supply.

2.2 Presenting the results of the risk assessment Using risk matrices

Consider the differential risk *r*(*Skl*) associated with the realisation of a particular accident scenario *Skl*, which is characterised by an adverse initiating impact *Hk*, causing the achievement of a particular unfavourable state of the KCI object, expressed as

$$r(S_{kl})=P(H_r)*P(KC_l|H_k)\cdot U(KC_1)$$

where $P(H_k)$ is the hazard defined as the probability of realisation of an adverse initiating impact (event) H_k ; the object reaching an undesirable CC state$_l$ in case of an adverse impact H_k ; U KC_l $\equiv U$ S_{kl} - damage when the object reaches an undesirable CC state, identically equal to the damage from the accident scenario S_{kl} .

Thus, the results of hazard and vulnerability analysis make it possible to estimate the probability of realisation of the complex event α associated with the realisation of the scenario S_{kl} $\equiv S_{kl}$ which is that an adverse impact *Hk* occurs and the object reaches a damaged state KS_l : S_{kl} $Hk \cap _l$ KC. The probability of a complex adverse event of a complex adverse event α is identically equal to the consequences in case the system reaches the final state of KS_l : $C \equiv CS_{kl}$ CKC $_l$

The expression for differential risk can be written as the product of two factors - the probability of an adverse event and its consequences (see (4.1))

$$r(S_{kl})\ P(S_{kl})\ X\Sigma\kappa\lambda$$

or

$$r()\ \Pi()\ X$$

The form of recording presented allows for a risk assessment approach based on the use of so-called risk matrices.

In order to systematise and analyse the resulting probability and consequence estimates, the probability and consequence scales are divided into intervals, which are assigned categories in Table 2.3.

Table 2.3. Probability-Heaviness of Consequences Matrix

Failure scenario		**Consequences of a failure scenario**			
		Neglected small	**Non-critical**	**Critical**	**Catastrophic**
		$<10^3$	$10^3 \dots 10^5$	$10^5 \dots 10^6$	$>10^6$
Probability	Frequent (>1 per year)	*R=2*	*R=3*	*R=4*	*R=4*
	Likely ($1 \dots 10^{-2}$ per year)	*R=1*	*R=2*	*R=3*	*R=4*
	Possible ($10^{-2} \dots 10^{-4}$ per year)	*R=1*	*R=2*	*R=3*	*R=3*
	Rare ($10^{-4} \dots 10^{-6}$ per year)	*R=1*	*R=1*	*R=2*	*R=3*
	Almost unbelievable ($<10^{-6}$ per year)	*R=1*	*R=1*	*R=1*	*R=2*

The grid divides the first quadrant of the probability-impact plane into a set of cells, each of which is assigned a certain level of criticality, also referred to with certain assumptions as risk level *R*. The resulting probability-impact matrix is usually called a risk matrix for short, although this name should be used with some caution, as the possibility of risk assessment using such matrices needs to be confirmed.

The criticality (risk) categories shown in Table 2.3. correspond to the following measures: *1* - risk not taken into account - no special (additional) safety measures need to be analysed and taken; *2* - acceptable risk - a qualitative hazard analysis or some protective measures are recommended; *3* - undesirable risk - a quantitative risk analysis is advisable or some safety measures need to be taken; *4* - unacceptable risk - a quantitative risk analysis is mandatory and some protective measures need to be taken.

Thus, the risk matrix is a table, of dimension *m* x *n*, consisting of *m* rows corresponding to the probabilities of various hazardous events (extreme natural events, failures of system elements or terrorist attacks on the system), and *n* columns corresponding to the different scales of consequences of these events.

The matrix rows are formed by dividing the consequence scale into *n* qualitative categories: (e.g. "minor", "moderate", "severe", "catastrophic"). At the same time, the consequences x of each adverse event α will fall into one of these *n* consequence categories (e.g. C"severe"). The columns of the matrix are defined by dividing the probability scale into *m* qualitative categories: (e.g. "almost improbable", "unlikely", "quite probable", etc.). Moreover, the probability P_α of each adverse event α will be assigned to one of the *m* categories listed above (e.g. P"quite probable").

Thus, each cell δ_{ij} of the risk matrix is defined by a pair of qualitative values of probability P_i and consequences C_j and defines one of s qualitative risk levels $R = l$, $l = 1, 2, ..., s$ for all points belonging to this cell. This level specifies the risk reduction (safety) measures.

Thus, each adverse event α will correspond to a pair of qualitative values on the probability and consequence scales (P_α; C_α), allowing to assign this event to a certain cell δ_{ij} of the risk matrix, which is assigned a certain risk level $R(\delta_{ij}) = l$, $l = 1, 2, ..., s$. For example, realization of the event connected with loss of tightness of a pressure vessel can be assigned to a pair of values (P_α = "probable", C_α = "severe") that causes this event to fall into a cell to which the experts who formed the risk matrix have assigned a risk level R = "high". Further, as the level of risk is deemed unacceptable, measures are selected to reduce the level of risk.

If the risk matrices are constructed correctly, the requirement of so-called *weak consistency* with the results of the quantitative risk assessment must be fulfilled. This means that if events α and β correspond to quantitative risks $r_\alpha = p_\alpha \cdot c_\alpha$ and $r_\beta = p_\beta \cdot c_\beta$, with $r_\alpha \geq r_\beta$, then the risk matrix must be constructed so that cell δ_A, in which event α falls, is assigned a risk level $R(\delta_A)$ that is not lower than the risk level $R(\delta_B)$ assigned to cell δ_B, in which event β falls: $R(\delta_A) \geq R(\delta_B)$. For example, if event α falls into a cell that is assigned a "medium" risk level, the cell into which event β falls should be assigned a "medium" or "low" (but not "high") level. In this case, the risk matrix allows us to distinguish between high and low risks with a certain degree of confidence. That is, from the condition $r_\alpha \geq r_\beta$ it follows that $R(\delta_A) \geq R(\delta_B)$. If this requirement is not met, then an event α, which is assigned a lower risk rating than an event β, may in fact correspond to a higher quantitative risk $r_\alpha > r_\beta$.

The analysis shows that the weak consistency requirement, is a rather strong constraint when constructing risk matrices. Two lemmas are proved in the paper [17]:

Lemma 1: If the matrix satisfies the weak consistency requirement: then cells that are assigned a risk level of 'high' should not border cells that are assigned a risk level of 'low'.

Lemma 2: None of the cells located in the leftmost column or the lowest row can be assigned a maximum risk level.

It follows that it is difficult to assess the risks of extreme events (events with low probabilities of occurrence and extremely high damages) using risk matrices.

The hypothesis that the risk matrix is an approximate qualitative representation of the original (possibly unknown) quantitative risk scale (Figure 2.5) suggests that arbitrarily small increments in probability and consequence size should not lead to jumps in risk prioritisation from the lowest to the highest level of risk without crossing an intermediate level.

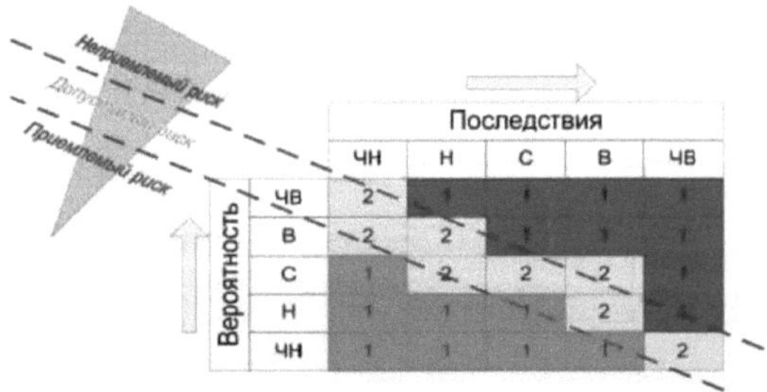

Figure 2.6. - Formation of a risk matrix based on the quantitative limits of the risk scale (WH - extremely low, L - low, S - medium, H - high, HH - extremely high)

Indeed, if the increasing levels of risk in the risk matrix represent (at least approximately) successive intervals of some original quantitative risk scale, then, with a continuous increase in risk from 0 to the maximum value, the corresponding qualitative levels of risk must also consistently increase. There should be at least one intermediate level between the minimum and maximum qualitative levels.

Thus, another requirement for the construction of risk matrices can be formulated, called the intermediate level requirement: The risk matrix must be formed so that any straight line that has a positive gradient and that starts in a cell with a "low" level of risk and ends in a cell assigned a "high" level of risk passes through one (or more) cells that have an "intermediate" (or more intermediate) level of risk.

However, it should be noted that the theoretical justification for using risk matrices as a risk comparison tool for rail transport facilities requires further development. There are a number of difficulties associated with this approach:

Firstly, the scales used are ordinal. Since many mathematical operations, in particular addition and multiplication, are not defined on ordinal scales, problems inevitably arise when trying to assess the level of security of a facility as a whole using risk matrices. Indeed, above we discussed the assessment of differential risks associated with the realisation of individual adverse events in the facility in question. Complex systems are characterised by the existence of multiple failure scenarios, realised with probability p_i, and having consequences c_i. In order to make a decision on the possibility of operating the facility as a whole, it is necessary, using quantitative risk analysis methods, to estimate the value of integral risk $r_{\Sigma} = \sum_i p_i * c_i$ and compare the received value with the maximum admissible value of risk g_{dop}. Obviously, it turns out to be impossible to solve the posed task by means of risk matrices, as risk matrices permit to receive only a qualitative assessment of levels of differential risks R_i, connected with realisation of separate failures. However, the obtained risk levels cannot be summed up in order to estimate the level of integral risk.

It should be noted that risk matrices assume an estimate for an adverse event based on a pair of values ("probability"; "consequences") and a ranking of the risk associated with that event, but do not answer the question of the cost of risk reduction measures.

In addition, it has been shown in [17-19] that the use of risk matrices is not always conducive to the implementation of risk-based management principles. Often the use of this approach results in completely different risks being grouped into a single cell. It should also be emphasised that risk matrices quite often produce results that are inconsistent with the results of quantitative risk assessment, as only the mean values (mat expectations) of the probability and consequence distributions are considered and the variance of these quantities is not taken into account.

The assignment of risk levels is highly subjective and reflects the perception of risk by the expert who forms the matrix or safety attitude adopted by the company operating the facility. Examples of conservative and risk-based approaches to safety assessment and management are presented in Figures 2.6(a) and (b), where R = 1 - low risk, no protective measures are required; R = 2 - medium risk, protective measures are implemented where possible, but activities may continue; R = 3 - high risk, urgent implementation of protective measures is required or operation of the facility must be suspended.

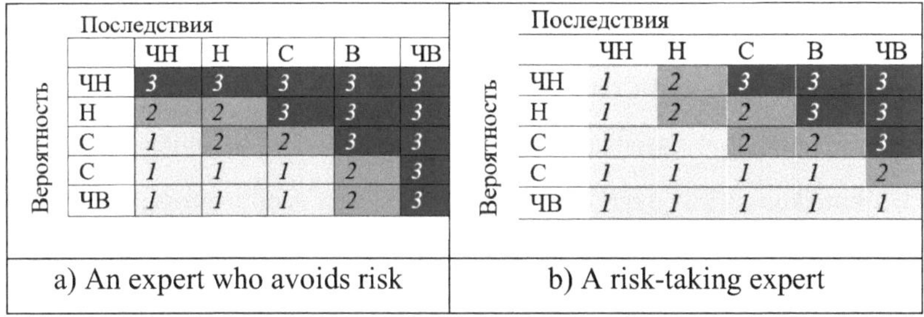

Вероятность \ Последствия	ЧН	Н	С	В	ЧВ
ЧН	3	3	3	3	3
Н	2	2	3	3	3
С	1	2	2	3	3
С	1	1	1	2	3
ЧВ	1	1	1	2	3

a) An expert who avoids risk

Вероятность \ Последствия	ЧН	Н	С	В	ЧВ
ЧН	1	2	3	3	3
Н	1	2	2	3	3
С	1	1	2	2	3
С	1	1	1	1	2
ЧВ	1	1	1	1	1

b) A risk-taking expert

Figure 2.6. - Examples of conservative and risk-based Approaches to safety assessment and management

However, the construction of risk matrices is now one of the most common and accepted methods of assessing safety and developing risk reduction measures.

The accuracy of the representation of risk and the reliability of the conclusions derived from risk matrices depends on the mutual distribution of probability and consequence values.

The simplest risk matrix (Figure 2.6) has a dimension of 2 x 2. Let the two risk factors - probability P and consequence C - take values in the range 0 to 1 (here damage values are normalised by degrees of system damage, with $C = 0$ corresponding to an intact system, and $C = 1$ corresponding to total system failure or maximum severe consequences). Each pair (P; C) is assigned a quantitative risk value $R = P \cdot C$. In the formation of the risk matrix, the boundary of areas between low and high values *of P* and C must be chosen. Let $x \in$ 0;1- a boundary value between low and high probability of an adverse event in the system in question, and $y \in$ 0;1- a value that is the boundary between low and severe consequences.

In order to assess the effectiveness of risk matrices as a decision-support tool for implementing protection measures, consider the following problem. Let the available resources be sufficient to handle only one of the risks. The decision maker must determine which of the two risks (α or β) should be handled first.

The random variables P and C are assumed to be independent and distributed uniformly over the interval [0; 1]. There is only information about which cell of the risk matrix each of the risks belongs to. It is necessary to determine which of the risks is quantitatively the greatest. The objective is to assess whether the qualitative categorisation of quantitative risks that is carried out using the risk matrix can be used to determine a solution that minimises the expected damages.

Table 2.4 - Simplest risk matrix

		Consequences	
Probability		Light	Heavy
	High	R = "Average"	R = "High"
	Low	R = "Low"	R = "Average"

The risks of events α and β can be ranked without error if the risk of one event falls in the "low" risk cell and the risk of the other in the "high" risk cell, because in this case any risk from the "high" cell is both qualitatively and quantitatively greater than the risk from the "low" cell. The probability of such an event is 2-(1 - x)-(1 - y)-x-y. This function reaches a maximum of 0.125 when x = y = 0.5. Otherwise, when two risks have the same qualitative rating, it is not possible to use the risk matrix to select the risk to be minimised in the first place. In this case, the probability of making an error can be assumed to be 0.5.

As will be shown below (see Lemma 1), where one of the two levels of risk belongs to the 'medium' cell and the other to the 'low' or 'high' cell, there is also a probability of error because some points in the cell with the higher quality rating have lower quantitative risk values than some points in the cell with the higher quality rating. The probability that two risks can be correctly ranked is 1∴2*1∴4 =0.125. This occurs when one falls into one cell of the diagonal 0;0,1;1 and the other falls into another cell of the same diagonal. The probability that the two risks cannot be ranked correctly using the risk matrix is

The probability that two risks can be ranked with a probability of error greater than 0 but less than 50% is 1 - 0.125 - 0.375 = 0.5.

Consider the more general case. Let $x = y = 0.5$, but the random variables P and C are correlated. If P and C are evenly distributed along the diagonal 0;0,1;1, then the probability that the two risks will be categorised without error (when one of the risks appears in the top right cell of the diagonal and the other appears in the bottom left cell of the same diagonal) is 50%. Otherwise, when both risks end up in the same cell, the probability of no error is also 50%. Thus, if P and C are positively correlated, the probability of error is 0.5-0.5 = 0.25.

In contrast, where the P and C values are negatively correlated and distributed along the diagonal 0;1,1;0, then both risks will be ranked as 'average' (even though their numerical values range from 0 at the ends of the diagonal 0;1,1;0и to 0.25 in the middle of the same diagonal). The risk matrix will not, however, provide useful information regarding the ranking of risks. That is, under these (less favourable) conditions, the risk matrix offers no advantage over random selection of the priority risk and the probability of error increases to 0.5.

The example shows that risk matrices can be useful in categorising risks and prioritising protective actions if the random variables of probability and consequence are positively correlated (Figure 2.7(a)). Such distributions occur, for example, in terrorist impacts where there is a positive relationship between the consequences of a terrorist attack and the probability of its occurrence.

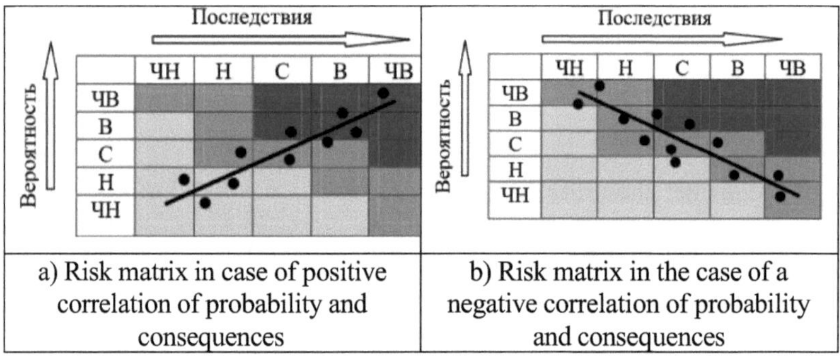

a) Risk matrix in case of positive correlation of probability and consequences

b) Risk matrix in the case of a negative correlation of probability and consequences

Figure 2.7. - Using the risk matrix for different correlations of probability and consequences

However, estimates and decisions made using risk matrices are often flawed when there is a negative correlation between the probability and consequences of a hazardous event (Figure 2.7(b)). Unfortunately, this is often encountered when considering accidents in technical systems, where both low probability and severe consequence failure scenarios and high probability and low consequence scenarios are characteristic.

2.3 Risk treatment phase

Risk treatment is an iterative process and involves

- choosing one or more options for risk reduction measures;
- planning of risk reduction measures;

- carrying out risk-reduction activities.

A flowchart explaining the actions in the risk treatment phase is shown in Figure 2.8

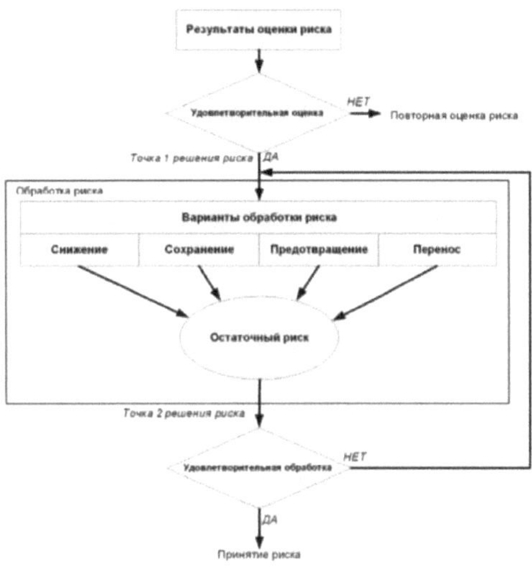

Figure 2.8. - Risk handling diagram

The results of the risk assessment phase provide recommendations for risk reduction, which are inputs to the risk treatment phase. Risk mitigation measures may be ***technical*** and/or ***organisational*** in nature.

In practice, the following four options for risk treatment measures are commonly used [20]:

- risk prevention - consideration of ways to eliminate the hazard or vulnerability, or to change a process or activity so that the hazard is no longer applicable to it. When identified risks are deemed too high, a decision may be made to stop or abandon a planned, or existing, activity altogether;

- Risk transfer - the transfer of risk to a third party who can take over the risk, such as insurance companies, or through the transfer of functions to network solution providers or security management services, outsourcing.

Risk transfer can create new risks or modify existing identified risks, so additional risk treatment may be necessary;

- Risk reduction - the application of appropriate controls for dangerous failures and other undesirable events reduces the frequency (likelihood) or size of possible consequences. Each control can provide one or more types of protection: prevention, containment, detection, mitigation, remediation, correction, monitoring and information;

- Risk acceptance - deciding on all remaining risk. The organisation must come to a decision to accept the risk based on the acceptance criteria. This decision arises for two reasons. The first reason is to successfully mitigate the risk when the residual risk after the implementation of controls does not exceed the risk acceptance criteria. The second reason is to retain the risk, i.e. even if the initial or residual risk exceeds the risk acceptance criteria,

the organisation's management makes the decision to accept the risk, taking into account various conditions, such as budget, time constraints, etc.

Risk treatment should be clearly prioritised, within each of which individual risk treatment options will be implemented. Prioritisation can be done using a variety of methods, including risk ranking and cost-benefit analysis.

As a result of the risk treatment, there is residual risk, the decision to accept it usually requires an assessment of it.

2.4 Risk monitoring and review phase

The main purpose of risk monitoring is to reduce uncertainty in risk assessment. The use of risk information in risk management decision-making brings risk monitoring into the risk management loop, which allows the risk monitoring process to be considered as an integral part of the risk management decision-making process.

Risk is also monitored in the process of controlling the effectiveness of risk management decisions.

Thus, functionally, risk monitoring is closely linked to both the analysis and evaluation functions as well as the risk management functions.

Risk monitoring activities should be repeated regularly, periodically, as the factors affecting risk may be constantly changing. the frequency of monitoring is determined separately for each type of risk, depending on its significance.

The risk monitoring tasks in Table 2.5 can be roughly divided into three main groups:

- analytical monitoring tasks;
- situational monitoring tasks;
- operational monitoring tasks.

Table 2.5. -Risk monitoring tasks

<table>
<tr><th colspan="3">Risk monitoring</th></tr>
<tr><th>Analytical monitoring</th><th>Situation monitoring</th><th>Operational monitoring</th></tr>
<tr><td>Highlighting risk indicators</td><td>Observation of sources of information on risk</td><td rowspan="3">Controlling the results
risk management</td></tr>
<tr><td>Assessing the magnitude of the risk</td><td>Tracking the dynamics of risk</td></tr>
<tr><td>Risk assessment</td><td>Control of parameters affecting risk</td></tr>
<tr><td colspan="3">Forming a database of the risk area of concern</td></tr>
</table>

Functionally, **analytical risk monitoring** can be represented as a step-by-step procedure for carrying out various operations (collection, processing, analysis, presentation and others):

(a) A list of possible risks is identified;

b) Identifying possible sources of information to monitor risks according to the list of risks identified and structured in the previous step;

c) gathering information from sources of risk information;

d) presenting the collected information in predetermined display forms for further processing and analysis;

e) Data processing and analysis using selected risk analysis methods. Various methods are used at this stage: modelling of processes whose development may give rise to risk; expert judgement; fuzzy knowledge processing; identification of hidden hazards and others;

(e) Based on the results of data processing and analysis, risk indicators, such as the occurrence of new hazards, changes in the frequency of occurrence and severity of consequences for already known hazards, etc., are highlighted.

g) a risk assessment is carried out based on the identified attributes in accordance with the adopted risk criteria.

i) Analyse the quality of risk monitoring of the type under consideration and make recommendations on risk monitoring quality management;

(j) Adjustments are made to the risk monitoring technology in the light of the recommendations made.

Situational risk monitoring is designed to identify, track and monitor informative signs of risk in a given situation.

Situational monitoring, as well as analytical monitoring, can be represented as a sequence of operations for processing different types of information, performed as a step-by-step procedure. Given the similarity of the semantic description of the first five technological operations (a-e) of analytical and situational risk monitoring, let us consider the step-by-step procedure of situational monitoring, starting from the sixth step (f):

(e) Modelling risk development processes over a given time horizon, taking into account the evolution of indicators in response to changes in the situation;

g) rapid risk assessment using a risk assessment decision support system;

i) output of the risk assessment data into the risk management decision-making system;

j) analysing the risk monitoring work carried out and making recommendations to improve the situation monitoring technology.

Operational risk monitoring is designed to address the task of monitoring risk during and as a result of risk management.

In addressing risk management, operational monitoring should be viewed as one of the functional components of risk management.

Technologically, operational risk monitoring is an element of the risk management decision support system.

Operational risk monitoring has two main tasks:

- monitoring risk in the management of a particular type of risk;

- analysing the effectiveness of risk management.

The decision support system for risk management addresses these issues.

Risks are not static. Risks, their occurrence, frequency of occurrence and consequences can change significantly as a result of various factors. Risk monitoring, carried out at different stages of the risk management process, is necessary to detect and control these changes. Risk monitoring can result in a recommendation to revise the risk.

If no special measures are taken, at some point the risk assessments previously obtained will be incorrect because they will not take account of changes in influencing factors that have occurred. For example, new hazards, changes in their likelihood or consequences may significantly increase the risks previously assessed as low.

A risk review, usually involving the risk analysis, risk assessment and risk treatment stages, is carried out:

(a) In identifying any changes in hazards, their manifestations, their frequency of occurrence and the severity of their consequences;

b) when significant operational, technical, economic, regulatory, social and environmental changes occur.

2.5 The relationship of proto-industry to other sciences and subjects

JSC "AGMK" is a plant with prospects of copper ore production growth due to its own deposits. In 2003, AGMK increased its copper ore production by 2.7% to 27.33 mln tons (Fig. 2.9). The dynamics of AGMK copper ore production in 2000-2003 period is stable. According to the mill's forecasts for 2004, the copper ore production is expected to remain practically at the level of 2003.

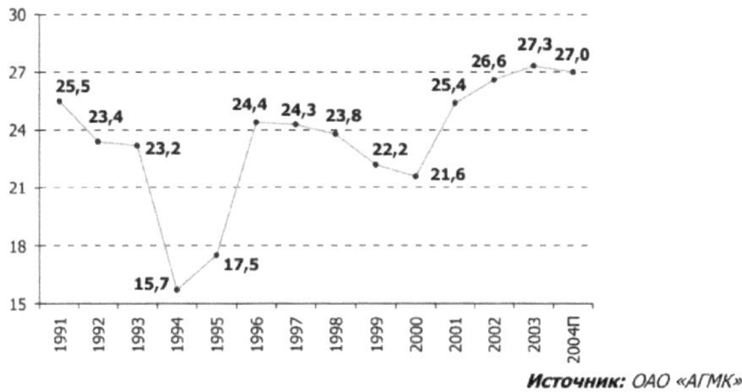

Fig. 2.9. Dynamics of AGMK's copper ore production growth, mln tonnes

As noted, copper ore is mined at AGMK's two major deposits, Kalmakyr and Sary-Cheku, the largest of which is Kalmakyr, which accounts for 92% (in 2003) of the mill's total ore output.

Despite the increase in copper ore production for the mill as a whole, it should be noted that the growth is mainly due to increased production from the Kalmakyr mine.

Ore processing at AGMK is carried out at the following concentrators: copper concentrator; lead concentrator. The smelting shop is equipped with a reflector furnace with an annual capacity of 50 thousand tonnes of blister copper, a flame furnace with a capacity of 65 thousand tonnes, four horizontal rotary converters with 75 tonnes capacity each and anode rotary furnaces with 200 tonnes capacity each; a copper electrolytic cell is equipped with electrolytic baths with an annual capacity of 147 thousand tonnes of copper cathode. this facility is equipped with electrolytic baths with an annual capacity of 147,000 tonnes of copper cathode per year, a gold and silver refinery (as well as selenium and tellurium) and a sludge and cupreous facility with an annual capacity of 7,000 tonnes of copper sulphate. Industrial accident - a hazardous accident in a shop resulting in damage to one or more furnaces to the extent of overhaul and (or) death of one or more persons, infliction of bodily injuries of varying severity to victims or complete interruption of the technological process in the emergency area exceeding the normative time. In disasters, the consequences are much more severe.

Emergencies, accidents and disasters, natural disasters (floods, landslides, hurricanes, fires, etc.) can cause accidents (disasters) of both types and cause loss of life at metallurgical facilities. In the area of earthquakes, metallurgical industry facilities are subject to destruction, damage and collapse. To protect such facilities from natural hazards, appropriate engineering structures are used: special galleries and retaining walls, drainage and bank protection structures (ditches, dams, traverses, etc.).

Safety of metallurgical industry facilities - a state of process safety against operational accidents, ensuring the safety of raw materials and end products, safety of personnel, preservation of the natural environment and the smooth functioning of the process.

Any autonomous system, in order to exist stably, needs subsystems that are responsible for protecting the safety and security of its own existence and that of its individual components. It can be said that the consistency of such subsystems and mechanisms also determines the stability of the system itself. Today, there are a number of branches of knowledge dealing with the protection of various aspects of human life. Both physical laws and special dependencies derived from them, which describe changes in the properties and state of materials, can be divided into two main groups. Firstly, these are the laws that describe the relationship between reversible processes, where, after the cessation of the action of external factors, the material (and, accordingly

part) returns to its initial state. These dependencies are called laws of state. Secondly, there are laws that describe irreversible processes and therefore make it possible to estimate those changes in the initial properties of materials that occur or may occur during the operation of the product. These dependencies are called ageing laws. Laws of state can be divided into static, where time is not a factor in the functional relationship between input and output parameters, and into transients, where changes in output parameters over time are taken into account. Typical examples of static laws of state are Hooke's law, the law of thermal expansion of solids, etc. These laws are used to derive computational dependencies for solving various engineering problems. Static laws of state, although not including the time factor, may be used in calculations of reliability if the characteristics of the product are known to change during its operation. State laws describing transient processes, such as oscillations of elastic systems, heat transfer processes, etc., although include the time factor, but also do not take into account changes occurring during operation of products [15,16,17]. They usually belong to the category of fast running processes or medium speed processes. Only with a known change in the level of external influences they can be used to solve reliability problems.

These examples show that different branches of science, agencies and services are involved in the protection of human life as such. Each is concerned with protection in terms of its own interest. Very often in such cases the achievements of other sciences in the field of protection are overlooked. In addition, only a systematic approach to the study of protection (as a separate, independent field) can develop it into a methodology.

In order to address the issues in a holistic manner, a profound fundamental scientific basis is needed. It can be said that the problems of protection, including state bases, in emergencies need to be centrally and comprehensively studied and subsequently developed under the auspices of a single science. Just as interaction problems are now being studied by synergetics, protection problems should be studied and developed under the aegis. For this reason, it has become necessary to understand protection issues more thoroughly - for example, the study of the laws of ageing, which reveal the physical nature of the irreversible changes that occur in the materials of a product, is of fundamental importance for the evaluation of the loss of performance of the product. Although the laws of ageing are always related to the time factor, some of them do not involve time directly, because the resulting dependencies are found to be related to other factors (such as energy), which in turn manifest themselves over time. Such dependencies will be called *laws of transformation*. A typical example of transformation laws are dependencies describing corrosion processes. It is difficult to draw laws directly reflecting changes of corrosion rate in time: firstly, due to polyvariation of corrosion processes, when a large number of factors have a simultaneous and often opposite effect on damage intensity, and secondly, corrosion can be not only evenly distributed over metal surface (for example, in form of oxide film), but also have local nature (local corrosion) or appear in form of intercrystalline corrosion. The laws of chemical thermodynamics are used to assess the possibility and intensity of the corrosion process. The application of physico-chemical laws to estimate the intensity of chemical corrosion processes is a typical approach to the analysis of complex phenomena of ageing and destruction of materials. Although it would be desirable to have a direct relationship between the time course of a given ageing process and the selection of optimal solutions, the complexity of the phenomenon does not allow this relationship to be obtained at this stage.

Therefore, physical and chemical laws are used, reflecting the most essential aspects of the process and indicators by which the intensity of the process can be indirectly judged. *Aging laws, which* evaluate the degree of damage to a material as a function of time, are the basis for solving reliability problems. They make it possible to predict the course of the ageing process, to estimate its possible realisations and to identify the most significant factors that influence the intensity of the process. A typical example of such dependencies are the laws of wear of materials, which on the basis of disclosure of physical picture of interaction of surfaces provide methods for calculation of wear intensity or wear value in function of time and estimate the parameters influencing the course of the process. Many temporal laws of physical-chemical processes can be derived on the basis of consideration of thermoactivation kinetics. Changes of properties of solids occur as a result of displacements and regroupings of elementary particles (atoms, molecules, electrons, protons, etc.) and changes of their positions in the crystal lattice. This refers to that small fraction of elementary particles, whose energy exceeds a certain level, which is called the activation energy Ea. The greater the rate of the process, the greater the number of particles with an energy higher than the activation energy. Any ageing process arises and develops only under certain external conditions. To assess the possible types of damage to the materials of machine parts, it is necessary to establish the area of existence of the aging process and, first and foremost, the conditions under which it occurs. For a process to occur, a predetermined level of stresses, speeds, temperatures or other parameters must normally be exceeded. This initial level or threshold of sensitivity is particularly important for fast-acting aging processes where there is an intense avalanche-like development of the process after onset. Often the sensitivity threshold is associated with some energy level that determines the start of a given process. For example activation energy *Ea* defines an energy level from which the process of change in material properties can proceed. The energy concept is the basis of the theory of crack initiation in metallic structures at medium stresses remaining below the yield strength. Failures caused by common causes (multiple faults) A multiple fault is an event in which several elements fail for the same reason. Such causes may include the following:

– *equipment design faults* (faults not identified at the design stage which lead to failures due to interdependence between electrical and mechanical subsystems or elements of a redundant system);

– *Operating and maintenance errors* (incorrect adjustment or calibration, operator negligence, incorrect handling.

– *environmental influences* (dust, dirt, temperature, vibration, and extreme conditions of normal operation);

– *external catastrophic impacts* (natural external events such as floods, earthquakes, fires, hurricanes);

– *a common manufacturer* (redundant equipment or components supplied by the same manufacturer may have common design or manufacturing defects. For example, manufacturing defects may be caused by incorrect material selection, errors in installation diagrams, poor quality soldering, etc.);

– *common external power supply* (common power supply for main and standby equipment, redundant subsystems or elements);

– *incorrect functioning* (incorrectly selected set of measuring instruments or inadequately planned protection measures).

A number of examples of multiple failures in nuclear power plants are known. For example, some parallel connected spring relays failed at the same time and their failures were caused by a common cause; two valves were set in the wrong position due to incorrectly unlocked couplings during maintenance; several failures of the communication panel occurred due to destruction of a steam line.

In some cases, the common cause does not cause a complete failure of a redundant system (simultaneous failure of several nodes, i.e. the limiting case), but a less serious overall reduction in reliability, resulting in an increased probability of joint failure of system nodes. which is also a mechanism for protecting the effectiveness of emergency response units at hazardous industrial facilities.

All elements of a system or sub-system are constantly in a state of mutual coherence and unity. In a generalised approach to protection, we can propose to group everything that is protected under the general concept of "system". As different as the systems themselves may be, so different must be the methods and techniques of their protection, but the algorithm and methodology of protection can generally be brought, as it were, to a common denominator.

In a few words, let's focus on the formation of Protecto-Industries. Many distinguished minds who worked on the development of the tools and mechanisms of cognition for humanity and its problems in different countries have contributed greatly to the formation and development of protection as a science. These are the founders of modern concepts of physics, chemistry, biology, medicine, law and politics. It should be noted that any science is a mechanism of knowledge of reality or, as the Scottish scientist Minto William noted in his works, "an instrument of thinking". Protectoindustrialism, as well as its separate directions, is one of such tools. Arming mankind with this mechanism of cognition will enable us to look at many phenomena from the perspective of protection, to assess them, although in a completely different, unaccustomed to them, but nevertheless no less useful way.

In addition, it is fair to say that, in our republic, the current stage in the development of this branch of science has proved very useful for the development of a number of sciences.

In the process of growth and as scientific research differentiates, there is a need to build an overall picture of the connections of the individual disciplines. The need for a sound allocation of material resources, the organisation of a learning system for the proper formation of a scientific worldview, and the improvement of the apparatus of knowledge, requires a classification that satisfies the basic requirements of the scientific community.

The subject is the common proto-industrial protection laws for metallurgical industrial facilities. Their essence, purpose and subsequent development.

Speaking of the specific features of protoindustries, it is necessary to point out the method of countermeasures developed and implemented in the protection of metallurgical industrial systems.

The problem of classification is a pressing one in many branches of science, as it is a convenient material for methodological analysis. It clearly shows the logic of the development of the problems of science itself and the interaction of this science with others, its staggering, specific forms. In this case, it is necessary to understand the basics of classification, its criteria and attributes, which must be sought in practice. It is practice that provides the basis for the perception of individual branches of science as objects worthy of independent study. The emergence in various countries of the science of security as a separate science indicates that society is forced to create special mechanisms capable of increasing its stability and reliability of existence.

The safety of the technological process of arc furnaces is a complex issue which requires specific approaches for its solution. At present, there are features that require serious solutions, both in terms of technical and informational support. One of the most urgent of these is the presence of a large number of unrelated structures and systems that solve a narrow problem (Figure 2.10).

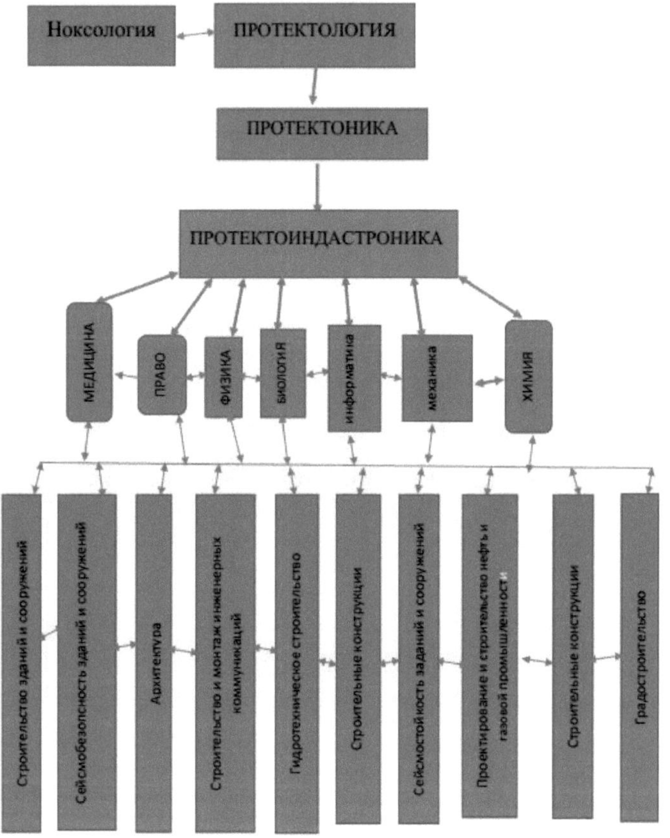

Figure 2.10 - Linking proto-industry with other sciences and subjects

This is due to historical factors. In the process of developing the process control system in the shops of MOF and COF, when solving individual tasks to ensure safety in one or another area of melting control, there was a need to use certain hardware and software automation tools. As a result, a large number of local solutions have accumulated, and the so-called "patchwork automation" has emerged.

A significant reserve for improving safety in batch plants is to combine individual safety subsystems into a single, multi-layered system. The integration should take place around the key element of the safety system - the reflection furnace itself. The multilevel control and safety system is supposed to be implemented as a set of three interacting hardware and software complexes.

Industrial accidents and catastrophes in hazardous industrial facilities are of two types: accidents (catastrophes) occurring at production facilities not directly related to the melting process (accidents at plants, stations, etc.), and accidents during the melting process.

In the area of earthquakes, metallurgical facilities are exposed to destruction, damage and collapse. To protect such facilities from natural hazards, appropriate engineering structures are used: special galleries and retaining walls, drainage and shore protection structures (ditches, dams, traverses, etc.). Technical systems become interconnected only due to the presence of such an essential link as a human being.

Conclusion

The monograph deals with the issues of risk assessment at hazardous industrial facilities and infrastructure, including hazard analysis, vulnerability analysis and damage analysis. The classification and possibilities of using various methods used in international practice for risk assessment in technical systems of various branches of economy are given.

The conducted consideration of risk management issues has shown that in the design and implementation of concepts, strategies and programmes of technical development of WGHE it is necessary to assess the forecast efficiency of functioning and development of industry and infrastructure of the facility, as well as the estimated risks of adverse events accompanying its reform and functioning. At that, the actual efficiency of functioning will be determined by the difference between the economic effect and risks.

When planning an activity in an EPO, an assessment should be made of the positive effects of the activity and the level of risk for each stage of the planned activity's implementation.

Analysis of the levels of actual risks in the implementation of activities and their comparison with the levels of acceptable risks will make it possible to reasonably decide, using quantitative indicators, on the acceptability or, conversely, the unacceptability of the implementation of certain projects, on the direction of their revisions and adjustments leading to a reduction of risks.

In general, the choice of risk assessment methods is determined by the following main factors: the potential hazard of the object of risk analysis; the potential hazard and damage during the transition from normal (routine) conditions of operation of a complex system "human - infrastructure object or hazardous production facility - living environment" to emergency and catastrophic (non-standard) conditions; availability of initial deterministic or statistical information on the implementation of risks at the previous stages of functioning of the said systems

For risk analysis in complex systems, combined methods should generally be used, as well as various modifications of the methods discussed herein. However, different approaches can be used for each of the risk components when assessing the mutual influence of risks.

Quantitative methods of risk assessment are particularly necessary when the perceived severity and extent of damage are high. Quantitative methods allow you to compare alternative security measures and determine which one provides the best protection.

In cases where a full quantitative analysis is not possible due to lack of information (data) about the system, its operating conditions, possible failures (accidents), the influence of human factors, etc., a comparative quantitative or qualitative risk ranking by experts (experts well-informed in the field) may be effective under such circumstances.

At the stage of hazard identification and preliminary risk assessments, it is recommended to apply qualitative risk analysis and assessment methods based on established procedures, specific tools (questionnaires, forms, questionnaires, instructions) and the practical experience of the implementers.

The methods considered can be applied in isolation or in addition to each other, and qualitative analysis methods can include quantitative risk criteria (mainly based on expert judgement using, for example, a probability - severity of consequences matrix of hazard ranking). A full quantitative risk analysis should make use of the results of qualitative hazard analysis wherever possible.

The risk assessment procedure for railway transport facilities involves a sequential analysis of the hazards to which the facility in question is exposed, an analysis of the vulnerabilities of the facility in relation to the hazards identified, and an analysis of the damage from accidents realized when the facility was vulnerable to the effects of the hazards.

The merits of the risk matrix approach include its simplicity, clarity and applicability to a wide range of risk assessment disciplines. Risk matrices provide a holistic system for assessing and documenting differential risks and provide preliminary recommendations for prioritising measures to address different risks.

A risk matrix can be used to construct a qualitative approximation of some (usually unknown) quantitative risk scale. For this approximation to be correct, the construction of risk matrices must comply with a number of requirements, which in practice are sometimes quite difficult to meet. By means of a risk matrix it is possible to compare correctly only a part of possible pairs of risks of dangerous events as risks of two dangerous events having different quantitative values of risk, can fall into the same cell of a risk matrix.

The accuracy of risk assessment using risk matrices depends on the nature of the joint distribution of probability and consequence values. In the case of a negative correlation between these values, the accuracy of the assessment is significantly reduced and the use of risk matrices for risk ranking and prioritisation of protective measures becomes problematic. In addition, this approach does not allow taking into account the cost of protective measures and the effectiveness of measures aimed at reducing risks, i.e. does not allow the formation of an optimal strategy for protecting the object. As a disadvantage of risk matrices should also be noted the influence of subjectivity in the assignment of scales of probability and consequences by experts.

The protection of industrial and infrastructure facilities must be based on the management of risks associated with the construction and operation of these facilities and includes actions to reduce three risk factors: natural and manmade and terrorist hazards to which these facilities are exposed; the vulnerability of facilities to these hazards; the damage caused by hazards at the facilities.

List of references

1. ISO/IEC 31010:2009 Risk management - Risk assessment techniques. - 135 c.

2. IEC 62278:2002 (IEC 62278:2002) Railways. Specification and demonstration of reliability, availability, maintainability and safety (RAMS). - 104 c.

3. Vetoshkin A.G., Razhivina G.P. Life Safety: Assessment of Industrial Safety. - Penza: Publishing house of Penza State Architectural Academy, 2002. - 172 c.

4. N.A. Makhutov, V.P. Petrov, R.S. Akhmetkhanov, D.O. Reznikov et al. Risk Analysis and Safety Management (Methodological Recommendations). MOSCOW: MGF. "Znanie (in Russian) 2008. - 134 c.

5. Makhutov N.A., Petrov V.P., Akhmetkhanov R.S., Reznikov D.O. et al. Strength, resource, survivability and safety of machines. Librocom, 2008. - 565 c.

6. Makhutov N.A., Petrov V.P., Akhmetkhanov R.S., Reznikov D.O. et al. Human Factor in Security Problems. Moscow: MGF "Znanie", 2008.

7. Makhutov N.A., Petrov V.P., Reznikov D.O. Assessment of survivability of complex technical systems // Problems of safety and emergencies. - M.: VINITI, 2009, №3. - C. 47-66.

8. Makhutov N.A., Petrov V.P., Reznikov D.O., Kuksova V.I. Identification of defining parameters of threats, vulnerability and security of critical facilities in relation to prevailing threats of natural, man-made and terrorist nature // Problems of Security and Emergencies. - MOSCOW: VINITI, 2008, № 2. - C. 70-77.

9. Makhutov N.A., Petrov V.P., Reznikov D.O., Kuksova V.I. Ensuring the security of critical facilities based on reducing their vulnerability // Problems of Security and Emergencies. - M.: VINITI, 2009, NO.2. - C. 23-30.

10. Makhutov N.A., Reznikov D.O. Vulnerability assessment of technical systems and its place in risk analysis procedure // Problems of risk analysis. Vol. 5. - 2008. - № 3. - C. 76-89.

11. Methodical recommendations on damage assessment from accidents at hazardous production facilities. RD 03-496-02, M., 2002. -102 c.

12. Makhutov N.A., Reznikov D.O., Petrov V.P. Risk assessment of accidents at KVO in view of possibility of realization of extreme damages // Problems of safety and emergency situations. - M.: VINITI, 2008. - C. 45-51.

13. Shoigu S.K., Vladimirov V.A., Vorobyov Y.L., Shakhramanyan M.A. Safety of Russia. Protection of Population and Territories from Natural and Man-made Emergencies, Moscow State Pedagogical University 'Znanie', 1999. - 205 c.

14. Makhutov N.A. Strength and Safety. Fundamental and Applied Research. Novosibirsk. Nauka. 2008. - 522 c.

15. Baker J., Schubert M., Faber M. On the assessment of robustness. Structural Safety 30 (2008). Pp. 253-267.

16. Agarwal J. Structural Vulnerability and Robustness. Department of Civil Engineering, University of Bristol, Bristol BS8 1TR, UK. Structural Safety 15 (2008). Pp. 53-67.

17. L.Cox. What's wrong with risk matrices. Risk analysis. Vol. 28. N2, 2008. Pp. 497-511.

18. L.Cox. What's wrong with Hazard-Ranking Systems? Risk analysis. Vol. 29. N9, 2009. Pp. 940-948.

19. Cox, L. A. Jr., Babayev, D., & Huber,W. Some limitations of qualitative risk rating systems. Risk Analysis, 25(3), 2005. Pp. 651-662.

20. Varfolomeev A.A. Information Risk Management. - M.: RUDN, 2008. - 158 c.

21. Russia's security. Legal, socio-economic and scientific and technical aspects. Functioning and development of complex national economy, technical, energy, transport systems, communications and communication systems. Section One. Moscow: ICF Znanie, 1998. - 448 c.

22. Gorsky V.G., Motkin G.A., Petrunin V.A., Tereschenko G.F., Shatalov A.A., Shvetsova-Shilovskaya T.N. Scientific and methodological aspects of accident risk analysis. - Moscow: Economics and Informatics, 2002. - 260 c.

23. Makhutov N.A., Kryshevich O.V., Pereezdchikov I.V., Petrov V.P., Tartashov N.I. Features of application of methods for hazard analysis of man-machine-environment systems based on fuzzy sets // Problems of safety in emergency situations. Vol. 1, 2001. - C. 99-110.

24. Methodical instructions on carrying out risk analysis of hazardous production facilities RD 03-418-01. Approved by Resolution No. 30 of Gosgortechnadzor of the Russian Federation dated July 10, 2001.

25. Enterprise Standard. Methodological Guidelines for Risk Analysis of Hydraulic Structures.STP VNIIG 210.02.NT-04.

26. Akimov V. A., Lapin V. L. L., Popov V. M., Puchkov V. A., Tomakov V. I., Faleev M. I. Reliability of Technical Systems and Technogenic Risk. - Moscow: ZAO FID "Delovoy Express", 2002. - 368 c.

27. Samokhvalov V.P., Borisova D.A., Materikina S.S., Inchina E.V. Model of modern FMEA procedure // Mechanics and Engineering. - 2010. - №4. - C. 817-822.

Printed by Books on Demand GmbH, Norderstedt / Germany